SpringerBriefs in Applied Sciences and Technology

SpringerBriefs present concise summaries of cutting-edge research and practical applications across a wide spectrum of fields. Featuring compact volumes of 50 to 125 pages, the series covers a range of content from professional to academic.

Typical publications can be:

- A timely report of state-of-the art methods
- An introduction to or a manual for the application of mathematical or computer techniques
- A bridge between new research results, as published in journal articles
- A snapshot of a hot or emerging topic
- An in-depth case study
- A presentation of core concepts that students must understand in order to make independent contributions

SpringerBriefs are characterized by fast, global electronic dissemination, standard publishing contracts, standardized manuscript preparation and formatting guidelines, and expedited production schedules.

On the one hand, **SpringerBriefs in Applied Sciences and Technology** are devoted to the publication of fundamentals and applications within the different classical engineering disciplines as well as in interdisciplinary fields that recently emerged between these areas. On the other hand, as the boundary separating fundamental research and applied technology is more and more dissolving, this series is particularly open to trans-disciplinary topics between fundamental science and engineering.

Indexed by EI-Compendex, SCOPUS and Springerlink.

Azman Ismail · Fatin Nur Zulkipli ·
Zalizah Awang Long · Andreas Öchsner
Editors

Advances in Technology Transfer Through IoT and IT Solutions

Springer

Editors
Azman Ismail
Centre for Women Advancement
and Leadership, Malaysian Institute
of Marine Engineering Technology
Universiti Kuala Lumpur
Lumut, Perak, Malaysia

Zalizah Awang Long
Malaysian Institute of Information
Technology
Universiti Kuala Lumpur
Kuala Lumpur, Malaysia

Fatin Nur Zulkipli
School of Information Science, College
of Computing, Informatics and Media
Universiti Teknologi MARA
Machang, Malaysia

Andreas Öchsner
Faculty of Mechanical Engineering
Esslingen University of Applied Sciences
Esslingen, Germany

ISSN 2191-530X   ISSN 2191-5318 (electronic)
SpringerBriefs in Applied Sciences and Technology
ISBN 978-3-031-25177-1   ISBN 978-3-031-25178-8 (eBook)
https://doi.org/10.1007/978-3-031-25178-8

This Springer imprint is published by the registered company Springer Nature Switzerland AG
The registered company address is: Gewerbestrasse 11, 6330 Cham, Switzerland

Preface

This book describes common applied problems that are solved with the use of digital technology. The digital technology has simplified most of our daily activities. Technology has been improving our quality of life where human capability alone is insufficient enough to be utilised. For any challenging tasks, digital technology helps to solve it in very efficient ways and thousands of them are solved on a daily basis without much notice in the public. Software and IT technology let us to complete tasks in just a moment that took days without this technical support. In that sense, this volume presents a few examples on how software and IT-based solutions were successfully applied in solving actual engineering problems.

Lumut, Malaysia
Machang, Malaysia
Kuala Lumpur, Malaysia
Esslingen, Germany

Azman Ismail
Fatin Nur Zulkipli
Zalizah Awang Long
Andreas Öchsner

Contents

Review: Current Trends in Artificial Intelligence on Healthcare

Shamini Janasekaran, Anas Zeyad Yousef, Amares Singh, and Nashrah Hani Jamadon

Abstract Artificial intelligence (AI) is a technique developed for learning how to interpret data following the operation of the brain neural network sand which uses a number of knowledge layers—including equations, patterns, laws, deep learning, and cognitive computation. In the area of healthcare, the use of artificial intelligence will grow. Healthcare companies and life science organizations are now using various forms of AI. The main implementation areas include guidelines for diagnosis and treatment, patient care and adherence; these are assisted by the increasing proliferation of clinical data, which contributes to improvements of paradigms in healthcare. AI-friendly solutions can define significant raw data relationships and have implementation value in virtually all areas of medicine, including medication growth, clinical decisions, patient care, and financial and organizational decisions. Healthcare practitioners may overcome difficult problems which alone cannot be solved or which artificial intelligence takes a long time. Artificial intelligence can be a powerful asset for medical professionals, allowing them to make greater use of their capabilities and create health value.

Keywords Artificial intelligence · Healthcare · Clinical decision support · Machine learning

S. Janasekaran (✉) · A. Z. Yousef
Centre for Advanced Materials and Intelligent Manufacturing, Faculty of Engineering, Built Environment and IT, SEGi University, 47810 Petaling Jaya, Selangor, Malaysia
e-mail: shaminijanasekaran@segi.edu.my

A. Z. Yousef
e-mail: anas_z98@live.com

A. Singh
School of Computer Science and Engineering, Taylor's University, No. 1, Jalan Taylor's, Subang Jaya, Selangor, Malaysia
e-mail: amares.singh@taylors.edu.my

N. H. Jamadon
Department of Mechanical and Manufacturing Engineering, Faculty of Engineering and Built Environment, The National University of Malaysia, 43600 Bangi, Malaysia
e-mail: nashrahhani@ukm.edu.my

© The Author(s), under exclusive license to Springer Nature Switzerland AG 2023
A. Ismail et al. (eds.), *Advances in Technology Transfer Through IoT and IT Solutions*,
SpringerBriefs in Applied Sciences and Technology,
https://doi.org/10.1007/978-3-031-25178-8_1

1 Introduction

Artificial intelligence (AI) in healthcare has made massive advances in recent times, leading to research and discussion about the possibility of AI replacing human physicians in the future [1]. The truth is, human physicians will not be replaced, although in some practical areas of healthcare such as radiology, AI will certainly help clinicians to make informed clinical choices or even replace human decision [2]. The recent popular applications of AI in healthcare are made possible by the increasing abundance of healthcare data and the rapid growth of big data analytics tools [3]. Effective AI strategies, driven by relevant clinical questions, can uncover clinically relevant information found in an abundance of evidence, which in turn can support clinical decision-making [4]. The current situation of artificial intelligence in healthcare is discussed here, in addition to its prospects, starting with the reasons for applying artificial intelligence in healthcare to the data that are analyzed by Ai systems and the mechanisms that cause important clinical results by AI systems and the types of diseases that are treated by AI societies [5].

2 Artificial Intelligence Trends in the Healthcare

Artificial intelligence is a group of multiple technologies which are related to the healthcare field and support them widely [6]. Some of the special AI technologies of high importance for healthcare will be explained in this review paper.

2.1 Machine Learning Technology, Neural Networks, and Deep Learning

Machine learning is a mathematical methodology for data matching models and "learning" by data training models. One of the most common examples of artificial intelligence is machine learning. The center of many AI techniques is a wide-ranging technology, and there are many forms of it [7]. Precision medicine is the most common application of traditional machine learning in healthcare, where the prediction treatment protocols are effective on a patient on the basis of different patient attributes and the situation of treatment [8]. The large volumes of data for machine learning and precision medicine require a dataset of training for which the result factor is recognized; this is known as supervised learning [9]. One of the most complex types of machine learning is the neural network—the most significant application is a classification system, such as deciding if a patient is likely to have a certain disorder. Issues are seen in forms of inputs and outputs and factors or 'tools' values that bind inputs to outputs [10]. The way neurons interpret signals has been compared, but the relation to brain activity is comparatively small. Machine learning

Fig. 1 Application of deep learning in healthcare [14]

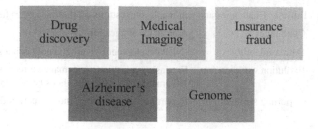

requires deep learning, with several layers of functionality or variables that forecast performance, or neural network models. In such models, there could be thousands of features concealed, which are exposed by the faster processing of GPUs and cloud architectures of technology [11]. Identifying suspected precancerous lesions on radiographs is a popular use of deep learning in healthcare. In oncology-oriented image processing, radiology and profound learning are generally seen. It seems that integrating them promises greater diagnostic precision than the previous generation of automatic image processing systems [12]. For speech recognition, deep learning is often increasingly used, and as such it is a type of natural language processing (NLP). Any function as of a deep learning model typically has no significance to the human observer, unlike previous modes of statistical analysis. As a result, decoding model findings can be very difficult or impossible to understand [13]. Figure 1 shows some of the applications that need deep learning in healthcare.

2.2 Natural Language Processing

Since the 1950s, learning human language has been a target of AI researchers. This area, natural language processing (NLP), involves language-related applications such as speech recognition, text processing, translation, and other objectives. Two fundamental approaches to this are NLP and statistical semantic. Statistical NLP depends on machine learning in particular, deep learning neural networks and has related to the recent improvement in the precision of recognition. It entails a "large group" or a huge contingent of languages to learn from [15]. In healthcare, the development, interpretation, and classification of clinical records and published research are predominant applications in NLP. NLP programmers can interpret unstructured medical clinical notes, generate information [16]. From the clinical viewpoint, study trials are generally modeled and analyzed at a patient or population level, such as forecasting how a group of patients may respond over time to particular therapies or patient surveillance. Although predictions at the person or group consumer level are considered by some NLP tasks, these tasks still constitute a minority [16]. Table 1 lists some natural language processing actionable suggestions.

Table 1 Challenges and actionable suggestions on the challenges faced [17]

Challenges	Actionable suggestions
Data availability	(NLP development) synthetic note generation, alternative governance
Evolution workbenches	Extrinsic evolution (performance effects in downstream task), Automated calibration methods (where is adaption needed)
Reporting standers	Report key development and evaluation details

2.3 Physical Robots

Robots execute fully functions in environments like factories and warehouses, such as moving, repositioning, welding or assembling items, and distributing medical supplies. Robots have been friendlier with humans in recent years and can be more effectively conditioned by pushing them through the appropriate mission [18]. They are getting smarter, as their "brains" integrate other AI capabilities (in fact, their operating systems). As time goes, the intelligence from other AI fields will be implemented into physical robotics [19]. Robotic surgery offers 'superpowers' to doctors, boosts their capacity to see, makes incisions that are minimally painful and minimally invasive, stitch wounds, etc. Even so, crucial choices are always taken by surgeons. Gynecological surgery, breast surgery, and head and neck surgery are typical surgical practices that use robotic surgery [20].

2.4 Robotic Process Automation

For administrative purposes, this technology executes organized digital functions, some requiring information structures, as if they were a human user following a script or rules. These are affordable, easy to the programmer and straightforward in their behavior, relative to other types of AI [21]. The automation of robotic processes (RPA) does not actually include robots, just computer programmers on servers. To function as a semi-intelligent user of the applications, it depends on a mixture of workflow, organization rules, and 'presentation layer' integration of information systems [22]. It is used in hospitals for routine procedures such as advance consent, hospital history updating, or billing. They can be used to retrieve data from, for example, faxed images and paired with other technology such as image recognition to input it into transactional structures [23].

3 Benefits of Artificial Intelligence in Healthcare

In the medical literature, the effects of AI have been extensively questioned. In order to read the attributes and characteristics of a large volume of healthcare data, AI implements complex algorithms, thus utilizing these observations and the data collected to assist clinical practice. Based on reviews, it can be fitted with learning and self-correcting capability to increase its performance [24]. By presenting the latest medical data from journals, manuals, and professional procedures to advise effective healthcare, an AI device may support clinicians [25]. In comparison, in human clinical practice, an AI system may help to reduce diagnosis and procedure mistakes that are unavoidable [26]. In addition, the AI system collects valuable knowledge from a vast variety of patients to help draw real-time conclusions on health risks and forecast health outcomes [27]. Table 2 shows the advantages of AI in healthcare for improvement in lifestyle.

3.1 Healthcare Data

In healthcare applications, AI systems should be equipped by data created from clinical practices, such as screening, diagnosis, and assignment of treatment to learn similar sets of subjects and correlations between subject characteristics and outcomes of interest [29]. In the form of demographics, patient reports, electronic recordings from medical instruments, physical tests, clinical labs, and images, such clinical data are also included but not limited [30]. Specifically, a significant majority of the AI literature analyses evidence from medical imaging, genetic testing, and electro-diagnosis at the diagnostic level [31]. The two primary types of data are physical examination notes and clinical laboratory results. The data are distinguished as they contain significant portions of unstructured narrative texts, such as clinical reports,

Table 2 Advantages of artificial intelligence in healthcare [28]

Benefits	Description
Fast and accurate diagnostic	AI helps in integrating information such as records with operating metrics which can assist physicians
Reduce human errors	Human errors may threaten patient safety due to lack of activeness. To overcome this AI as a superhuman spell checker will assist doctors by eliminating human errors
Cost reduction	With artificial intelligence the patient can get the doctor assistance without visiting hospitals which results to reduce the cost
Virtual presence	Using a remote presence robot, doctors can easily coordinate with their staff and patient in the hospitals and assist their queries

Fig. 2 Potential benefits of machine learning and AI in healthcare

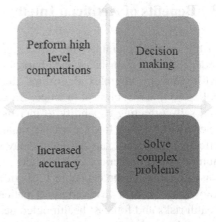

which cannot be specifically analyzed, with photographs, genetics, and electrophysiology details [32]. As a result, the subsequent AI programs concentrate on first translating the unstructured document into an integrated electronic detailed medical record (EMR). For instance, artificial intelligence tools have been used to derive phenotypic characteristics from case records to increase the accuracy of diagnosing congenital abnormalities [33]. Figure 2 shows the types of AI data systems [34] where AI and machine learning can potentially benefit humans.

4 Conclusion

There are still many problems surrounding the use of artificial intelligence in the healthcare industry, including some related to safety and those related to the poor efficiency or prejudice of the healthcare sector [35]. Nevertheless, in healthcare and disease diagnostics, AI certainly carries tremendous benefits, especially in high-disease or low-resource settings. It is still a medical instrument, though and is still far from hitting a point close to how human doctors do for this sort of job which is highly dependent on empathy and connection, which is what AI lacks. In the healthcare sector, empathy is an essential instrument [36]. In comparison, AI has no capacity to develop this relationship of empathy [37]. Therefore, artificial intelligence is unlikely to replace the doctor in diagnosing diseases any time soon, and instead, it will play a supportive role by providing signals and clues that help the doctor interpret the patient's symptoms, speed up the diagnosis process, and develop an early treatment plan [38].

Acknowledgements This research was supported by SEGi University, grant number: SEGiIRF/2022-Q1/FoEBEIT/001.

References

1. F. Jiang, Y. Jiang, H. Zhi, Y. Dong, H. Li, S. Ma, Y. Wang, Q. Dong, H. Shen, Y. Wang, Artificial intelligence in healthcare: past, present and future. Stroke Vasc. Neurol. **2**(4), 230–243 (2017)
2. A. Segato, A. Marzullo, F. Calimeri, E. De Momi, Artificial intelligence for brain diseases: a systematic review. APL Bioeng. **4**(4), 1–35 (2020)
3. A. Belle, R. Thiagarajan, S.M.R. Soroushmehr, F. Navidi, D.A. Beard, K. Najarian, Big data analytics in healthcare. Biomed. Res. Int. **370194**, 1–16 (2015)
4. T. Davenport, R. Kalakota, The potential for artificial intelligence in healthcare. Future Hosp. J. **6**(2), 94–98 (2019)
5. S. Sundvall, Artificial intelligence, in *Critical Terms in Futures Studies* (2019), pp. 29–34
6. J. Neves, H. Vicente, M. Esteves, F. Ferraz, A. Abelha, J. Machado, J. Machado, J. Neves, J. Ribeiro, L. Sampaio, A deep-big data approach to health care in the AI age. Mob. Netw. Appl. **23**, 1123–1128 (2018)
7. S. Wang, R.M. Summers, Machine learning and radiology. Med. Image Anal. **16**(5), 933–951 (2012)
8. C.C. Aggarwal, An introduction to neural networks, in *Neural Networks and Deep Learning* (2018), pp. 1–52
9. D. Ravi, C. Wong, F. Deligianni, M. Berthelot, J. Andreu-Perez, B. Lo, G.Z. Yang, Deep learning for health informatics. IEEE J. Biomed. Health Inform. **21**(1), 4–21 (2017)
10. A. Callahan, N.H. Shah, Machine learning in healthcare, in *Key Advances in Clinical Informatics: Transforming Health Care through Health Information Technology* (2017)
11. J. Schmidhuber, Deep learning in neural networks: an overview. Neural Netw. **61**, 85–117 (2015)
12. X. Glorot, Y. Bengio, Understanding the difficulty of training deep feedforward neural networks. J. Mach. Learn. Res. **9**, 249–256 (2010)
13. Y. Tsuruoka, Deep learning and natural language processing. Brain Nerve **71**(1), 45–55 (2019)
14. A. Kulkarni, A. Shivananda, Deep learning for NLP, in *Natural Language Processing Recipes* (2019), pp. 185–227
15. J. Hirschberg, C.D. Manning, Advances in natural language processing. Science **349**(6245), 261–266 (2015)
16. Y. Xie, L. Le, Y. Zhou, V.V. Raghavan, Deep learning for natural language processing, in *Handbook of Statistics*, vol. 38 (Elsevier, 2018), pp. 317–328
17. K. Kreimeyer, M. Foster, A. Pandey, N. Arya, G. Halford, S.F. Jones, T. Botsis, Natural language processing systems for capturing and standardizing unstructured clinical information: a systematic review. J. Biomed. **73**, 14–29 (2017)
18. L.D. Riek, Healthcare robotics. Commun. ACM **60**(11), 68–78 (2017)
19. J. Kim, G.M. Gu, P. Heo, Robotics for healthcare, in *Biomedical Engineering: Frontier Research and Converging Technologies* (Springer, Cham, 2016)
20. N. D'Elia, F. Vanetti, M. Cempini, G. Pasquini, A. Parri, M. Rabuffetti, M. Ferrarin, R. Molino Lova, N. Vitiello, Physical human-robot interaction of an active pelvis orthosis: toward ergonomic assessment of wearable robots. J. Neuroeng. Rehabil. **14**(1), 1–14 (2017)
21. W.M. Van der Aalst, M. Bichler, A. Heinzl, Robotic process automation. Bus. Inf. Syst. Eng. **60**(4), 269–272 (2018)
22. K.C. Moffitt, A.M. Rozario, M.A. Vasarhelyi, Robotic process automation for auditing. J. Emerg. Technol. Account. **15**(1), 1–10 (2018)
23. M. Lacity, L.P. Willcocks, A. Craig, Robotic process automation at Telefonica O2 (2015)
24. N. Parisis, Medical writing in the era of artificial intelligence. Med. Writ. **28**, 4–9 (2019)
25. M. Rowe, An introduction to machine learning for clinicians. Acad. Med. **94**(10), 1433–1436 (2019)
26. S. Reddy, J. Fox, M.P. Purohit, Artificial intelligence-enabled healthcare delivery. J. R. Soc. Med. **112**(1), 22–28 (2019)

27. P. Tschandl, C. Rinner, Z. Apalla, G. Argenziano, N. Codella, A. Halpern, M. Janda, A. Lallas, C. Longo, J. Malvehy, J. Paoli, S. Puig, C. Rosendahl, H.P. Soyer, I. Zalaudek, H. Kittler, Human–computer collaboration for skin cancer recognition. Nat. Med. **26**(8), 1229–1234 (2020)
28. K. Yeung, Recommendation of the council on artificial intelligence (OECD). Int. Leg. Mater. **59**(1), 27–34 (2020)
29. S.T. Liaw, H. Liyanage, C. Kuziemsky, A.L. Terry, R. Schreiber, J. Jonnagaddala, S. de Lusignan, Ethical use of electronic health record data and artificial intelligence: recommendations of the primary care informatics working group of the international medical informatics association. Yearb. Med. Inform. **29**(1), 051–057 (2020)
30. W. Nicholson Price II, Artificial intelligence in health care: applications and legal issues (2017)
31. A.L. Fogel, J.C. Kvedar, Artificial intelligence powers digital medicine. NPJ Digit. Med. **1**(1), 1–4 (2018)
32. J. Xu, K. Xue, K. Zhang, Current status and future trends of clinical diagnoses via image-based deep learning. Theranostics **9**(25), 7556 (2019)
33. K. Becker, J. Gottschlich, AI Programmer: autonomously creating software programs using genetic algorithms, in *Proceedings of the Genetic and Evolutionary Computation Conference Companion* (2021), pp. 1513–1521
34. S. Agarwal, S. Makkar, D.T. Tran, *Privacy Vulnerabilities and Data Security Challenges in the IoT* (CRC Press, 2020)
35. Y. Duan, J.S. Edwards, Y.K. Dwivedi, Artificial intelligence for decision making in the era of Big Data—evolution, challenges and research agenda. Int. J. Inf. Manage. **48**, 63–71 (2019)
36. Y.K. Dwivedi, L. Hughes, E. Ismagilova, G. Aarts, C. Coombs, T. Crick, M.D. Williams, Artificial intelligence (AI): multidisciplinary perspectives on emerging challenges, opportunities, and agenda for research, practice and policy. Int. J. Inf. Manage. **57**, 101994 (2021)
37. H.H. Haladjian, C. Montemayor, Artificial consciousness and the consciousness-attention dissociation. Conscious. Cogn. **45**, 210–225 (2016)
38. V.Y. Londhe, B. Bhasin, Artificial intelligence and its potential in oncology. Drug Discov. **24**(1), 228–232 (2019)

The Predictive Learning Analytics for Student Dropout Using Data Mining Technique: A Systematic Literature Review

Nurmalitasari, Zalizah Awang Long, and Mohammad Faizuddin Mohd Noor

Abstract This research aims to make a systematic review of the literature with the theme of predictive learning analytics (PLA) for student dropouts using data mining techniques. The method used in this systematic review research is the literature from empirical research regarding the prediction of dropping out of school. In this phase, a review protocol, selection requirements for potential studies, and methods for analyzing the content of the selected studies are provided. The PLA is a statistical analysis of current data and historical data derived from student learning processes to develop predictions for improving the quality of learning by identifying students who are at risk of failing in their studies. PLA in higher education (HE) is essential to improve knowledge. The failure of the HE to identify the potential factors contributing to student failure rate will risk both the HE images and the student's life. The systematic literature review conducted in this study was taken from selected journals published from 2016 to 2021.

Keywords PLA · Student dropout · Data mining · SLR

1 Introduction

Dropouts are a significant issue in many countries' higher education systems, as in United States [1]; Italy [2]; Europe [3]; Turkey [4], as well as Indonesia. Based on the Higher Education Data Base, Indonesia has increased students' percentage dropping

Nurmalitasari (✉) · Z. Awang Long · M. F. Mohd Noor
Malaysian Institute Information Technology, Universiti Kuala Lumpur, Kuala Lumpur, Malaysia
e-mail: nurmalitasari@s.unikl.edu.my

Z. Awang Long
e-mail: zalizah@unikl.edu.my

M. F. Mohd Noor
e-mail: mfaizuddin@unikl.edu.my

Nurmalitasari
Faculty of Computer Science, Duta Bangsa University of Surakarta, Surakarta, Indonesia

© The Author(s), under exclusive license to Springer Nature Switzerland AG 2023
A. Ismail et al. (eds.), *Advances in Technology Transfer Through IoT and IT Solutions*,
SpringerBriefs in Applied Sciences and Technology,
https://doi.org/10.1007/978-3-031-25178-8_2

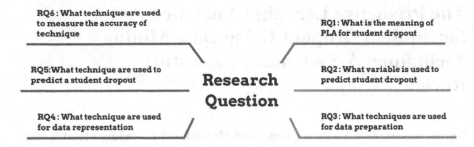

Fig. 1 The research question (RQ) of this systematic literature review

out over the last two years. Dropout has a tremendous negative impact on individuals, universities, and socioeconomics [5]. Many educational institutions need to prevent dropout students from considering the extent of the students' dropouts [6]. Early identification of at-risk students is critical for reducing dropout rates [7]. One way to see before on students' risk of failing in their studies is the predictive learning analysis (PLA) [8–10]. Nowadays, there are various techniques proposed to implement PLA. Data mining is one of the most famous analyzing student dropouts [11]. Data mining is used to identify patterns among large data sets and extract meaningful knowledge.

Research related to the systematic review of the literature on predicting dropout students using data mining techniques has been carried out by [12]. In this research of [12], there are 5 points discussed, namely: what techniques are used for data pre-processing, what factors affect dropout? what techniques are used for factor selection, what techniques are used to prediction and what are their levels of reliability, what tools are used? In contrast to [12], this study will discuss six research questions (RQ), as shown in Fig. 1.

2 Methodology

There are three primary elements of doing SLR in this study: (1) planning: this stage establishes the need for research and shows the review protocol; (2) conducting: this stage carries out the plan; the defined protocol is followed, as are the inclusion and exclusion criteria; and (3) reporting: this stage presents a statistical analysis of the selected documents and findings. Details of each stage are depicted in Fig. 2.

2.1 Planning Stage

The research question (RQ) was established to ensure that the review remained focused. Six research questions are posed to elicit information about the PLA for student dropouts developed using data mining techniques, which can be seen in

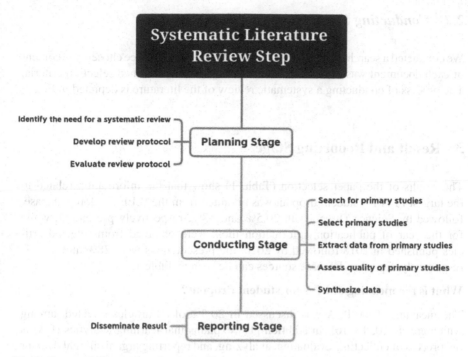

Fig. 2 Systematic literature review step

Fig. 1. The search process entails several steps, including selecting digital libraries, specifying search strings, conducting pilot searches, finetuning search strings, and retrieving a preliminary list of primary studies that match the search strings from digital libraries. The following databases were searched for articles from journals and conferences in SC imago Journal Country Rank (SJR) with an impact factor, such as ACM Digital Library, IEEE Explore, Science Direct, Springer, Taylor and France, Eric, Semantic Scholar, MDPI, Emerald, Google Scholar.

Finally, the following query was used:

"Predictive Learning Analytics" OR "Predict" OR "Prediction" AND "student dropout" OR "Student drop out" OR "Student dropping" AND "data mining"

The period of issue of the document being searched is between January 2016 and December 2021. Selection criteria for documents include (1) Research papers that address the research questions; (2) Papers that use data mining to predict student dropout rates; (3) Models to address the issue of university student dropout; (4) Documents describing the factors that contribute to university dropout; (5) Articles that discuss metrics for evaluating the performance of predictive models of dropout student.

2.2 Conducting Stage

We conducted a search using the strategy outlined in Sect. 2. Once chosen, the content of each document was examined to ensure that it met the defined selection criteria. The process of conducting a systematic review of the literature is depicted in Fig. 2.

3 Result and Reporting Stage

The results of the paper selection (Table 1) show that the information related to the largest PLA for student dropout was obtained from the IEEE Explore database, followed by Science Direct, with 20.5% and 18%, respectively presenting, while for the year of publication, information most were obtained from selected articles published in 2018 followed by 2020, the presentations were 28% and 20.5%, respectively, for other database sources can be seen in Table 1.

What is the meaning of PLA for student dropout?

The meaning of the PLA was discussed in 20.5% of all articles selected, among which are [8–10, 13–16]. In addition [17] explains that learning analytics (LA) is the process of collecting, evaluating, analyzing, and reporting organizational data for decision-making. Predictive learning analysis (PLA) is used to identify learners who may not complete their studies, who are usually described as being at risk [8–10, 14]. PLA can improve learning practice by transforming the ways teachers support the learning process [10, 13]. Meanwhile, according to [9, 17], PLA aims to improve learning by identifying students at risk of failing their studies.

In the PLA implementation, the teachers are the person who use PLA results to intervene and support students. The students are the individuals who received PLA

Table 1 Selected papers

Source	Identified paper	Selected paper
Semantic Scholar: https://www.semanticscholar.org/	361	3
ACM Digital Library: https://dl.acm.org/	3394	2
Science Direct: https://www.sciencedirect.com/	100.000+	7
Eric: https://eric.ed.gov/	13	2
Springer: http://link.springer.com	2304	4
MDPI: https://www.mdpi.com/	3	3
Emerald: https://www.emerald.com/insight/	278	1
Taylor & Francis: https://taylorandfrancis.com/	3249	3
IEEE Explore: https://ieeexplore.ieee.org/Xplore/home.jsp	796	8
Google Scholar: https://scholar.google.com/	2260	6
Total		39

interpretation from their teacher, and the education manager is tasked with evaluating and implementing PLA across universities [8–10, 17]. Predictive learning analytics aims to improve student performance by identifying students at risk of failing their studies [9]. Nonetheless, little is known about the most effective methods for integrating and scaffolding PLA initiatives within higher education institutions. To accomplish this, it becomes critical to collect and analyze the perceptions of relevant educational stakeholders (i.e., administrators, teachers, and students) regarding PLA [9]. It can then be concluded that PLA is a statistical analysis of current data and historical data derived from student learning processes to develop predictions of improving the quality of learning by identifying students who are at risk of failing in their studies. PLA in higher education is very important to improve learning. The failure for the HE identifying the potential factors contributing to student failure rate will risks both the HE images and student's life. The greatest risk of failure of the students in their studies is when the student is hit by a dropout.

What variable is used to predict student dropout?

Variables are constructs in research. In predicting dropout students, the variable is an important component for the success of the prediction. Based on a review of 39 journals (Table 1), 128 variables were obtained to predict student dropouts. These variables include cumulative grade point average (CGPA) [18–24]; gender [20–27]; age [20–23, 25–28]; father income [19, 22]; dropout indicator [21, 23–25, 27–29]; and others. These variables have linear and non-linear characteristics, low frequency, nominal, and ordinal measurement levels.

What technique is used for data preparation?

Data preparation is the process of transforming unstructured data into structured data. The purpose of this data preparation is to (1) Increase the effectiveness of PLA for dropout students; (2) Improve the data quality in order to improve the results of the PLA process; and (3) Simplify the information in order to facilitate the selection of appropriate techniques and PLA for student dropout analysis [30]. In this step data preparation only, those attributes were selected which were needed for data mining. For this, remove missing values; smoothing noisy data, selection of relevant attribute from database or removing irrelevant attributes, identifying or remove outlier values from data set, and resolving inconsistencies of data [19]. Not all articles use the data preparation stage, there are only a few who use it, and even then, it is not explained in detail at this stage. Of the selected articles, 28% articles use the data preparation process, namely, data cleaning [19, 22, 23, 26, 27, 31], data integration [21, 23, 26, 32], data reduction [21, 29, 33, 34], and data transformation [21].

What technique is used for data representation?

Data representation of dropout students is constantly increasing, electronic data collection is increasing, and various forms are available for data collection. Three data models are present, according to different studies: unstructured data, semi-structured data, and structured data. Of the selected articles, there are 20% articles that discuss data preparation, namely [23–28, 35, 36]. All data is represented in the form of Unstructured Data.

What technique is used to predict a student dropout?

The prediction is based on historical data in any data processing situation, with data mining techniques used to synthesize the relationships between variables. Based on the selection of 39 articles, 26 models were obtained to predict dropout students, including logistic regression [18, 21, 24, 26, 27, 32, 34, 36–43]; Decision tree [21, 23–25, 32, 37–40, 44–47]; Support Vector Machine [23, 26, 32, 37, 38, 45, 46, 48]; Naive Bayes (NB) [21–23, 25, 26, 34, 36], Random Forests [18, 21–24, 37–39, 49]; K-Nearest Neighbors [18, 23–25, 45, 46], Neural Network (NN) [24–26, 47], and others. The findings showed that logistic regression works very well in predicting dropout learners [18, 24, 36, 41–43].

What technique is used to measure the accuracy of technique?

Predictive models should be evaluated before being applied to predict dropouts. Inappropriate and inadequate model testing can result in inefficient models and incorrect predictions. There are several methods used to see the level of accuracy of the prediction model for dropout students from the selected articles, including ROC Curve [18, 21, 24, 25, 38, 47]; Accuracy [23–25, 28, 32, 33, 37–39, 49]; Precision [23, 34, 36, 40, 46, 47]; Recall [23, 34, 36, 38, 40, 46]; F-Measure [36, 38, 40]; TP Rate [27, 34, 47]; MSE [31]; MAE [31]; and Specificity [23]. The method that is often used to see the level of model accuracy is Accuracy. The selection of each method used to see the level of model accuracy is determined according to the prediction model used.

4   Conclusion

This study conducted a systematic literature review of the various aspects of data mining that have been considered for predicting university dropout. We identified 39 documents, which were selected according to the established inclusion and exclusion criteria, identifying six critical dimensions: (1) the meaning of PLA for student dropout; (2) variable is used to predict student dropout; (3) techniques are used for data preparation; (4) technique is used for data representation; (5) techniques are used to predict a student dropout; (6) techniques are used to measure the accuracy of the technique. The results of this analysis can be used as a basis for further research related to PLA for dropout students using data mining techniques.

References

1. Y. Chen, A. Johri, H. Rangwala, Running out of STEM: a comparative study across STEM majors of college students at-risk of dropping out early, in *ACM International Conference Proceeding Series* (2018), pp. 270–279
2. F. Jiménez, A. Paoletti, G. Sánchez, G. Sciavicco, Predicting the risk of academic dropout with temporal multi-objective optimization. IEEE Trans. Learn. Technol. **12**, 225–236 (2019). https://doi.org/10.1109/TLT.2019.2911070
3. S. Sultana, S. Khan, M.A. Abbas, Predicting performance of electrical engineering students using cognitive and non-cognitive features for identification of potential dropouts. Int. J. Electr. Eng. Educ. **54**, 105–118 (2017). https://doi.org/10.1177/0020720916688484
4. S. Aydin, A. Öztürk, G.T. Büyükköse et al., An investigation of drop-out in open and distance education. Educ. Sci. Theory Pract. **19**, 40–57 (2019). https://doi.org/10.12738/estp.2019.2.003
5. M.S. Larsen, K.P. Kornbeck, R.M. Kristensen et al., Dropout phenomena at universities: what is dropout? Why does dropout occur? What can be done by the universities to prevent or reduce it? A systematic review (2013)
6. B. Perez, C. Castellanos, D. Correal, Applying data mining techniques to predict student dropout: a case study, in *Proceedings of the 2018 IEEE 1st Colombian Conference on Applications in Computational Intelligence (ColCACI 2018)* (2018). https://doi.org/10.1109/ColCACI.2018.8484847
7. J.M. Ortiz-Lozano, A. Rua-Vieites, P. Bilbao-Calabuig, M. Casadesús-Fa, University student retention: best time and data to identify undergraduate students at risk of dropout. Innov. Educ. Teach. Int. **57**, 74–85 (2020). https://doi.org/10.1080/14703297.2018.1502090
8. C. Herodotou, Z. Zdrahal, B. Rienties et al., Implementing predictive learning analytics on a large scale: the teacher's perspective, in *ACM International Conference Proceeding Series* (2017), pp. 267–271. https://doi.org/10.1145/3027385.3027397
9. C. Herodotou, B. Rienties, B. Verdin, A. Boroowa, Predictive learning analytics "at scale": towards guidelines to successful implementation in higher education based on the case of the Open University UK. J. Learn. Anal. **6**, 85–95 (2019). https://doi.org/10.18608/jla.2019.61.5
10. C. Herodotou, B. Rienties, A. Boroowa et al., A large-scale implementation of predictive learning analytics in higher education: the teachers' role and perspective. Educ. Technol. Res. Dev. **67**, 1273–1306 (2019). https://doi.org/10.1007/s11423-019-09685-0
11. E. Iwatani, Overview of data mining's potential benefits and limitations in education research. Pract. Assess., Res. Eval. **23**, 1–8 (2018)
12. M. Alban, D. Mauricio, Predicting university dropout trough data mining: a systematic literature. Indian J. Sci. Technol. **12**, 1–12 (2019). https://doi.org/10.17485/ijst/2019/v12i4/139729
13. O. Viberg, M. Hatakka, O. Bälter, A. Mavroudi, The current landscape of learning analytics in higher education. Comput. Hum. Behav. **89**, 98–110 (2018). https://doi.org/10.1016/j.chb.2018.07.027
14. C. Herodotou, B. Rienties, M. Hlosta et al., The scalable implementation of predictive learning analytics at a distance learning university: insights from a longitudinal case study. Internet High. Educ. **45**, 1–13 (2020). https://doi.org/10.1016/j.iheduc.2020.100725
15. A.A. Mubarak, Predictive learning analytics using deep learning model in MOOCs' courses videos. Educ. Inf. Technol. (2020)
16. E.M. Queiroga, J.L. Lopes, K. Kappel et al., A learning analytics approach to identify students at risk of dropout: a case study with a technical distance education course. Appl. Sci. **10** (2020). https://doi.org/10.3390/app10113998
17. B.T.M. Wong, Learning analytics in higher education: an analysis of case studies. Asian Assoc. Open Univ. J. **12**, 21–40 (2017). https://doi.org/10.1108/aaouj-01-2017-0009
18. L. Aulck, N. Velagapudi, J. Blumenstock, J. West, Predicting student dropout in higher education, in *Machine Learning in within the Open Polytechnic of New Zealand, relying Social Good Applications* (2016), pp. 16–20

19. E. Balraj, D. Maalini, A survey on predicting student dropout analysis using data mining algorithms. Int. J. Pure Appl. Math. **118**, 621–627 (2018)
20. A. Pérez, U. Bío-bío, E.E. Grandón et al., Comparative analysis of prediction techniques to determine student dropout: logistic regression vs decision trees, in *37th International Conference of the Chilean Computer Science Society (SCCC)* (IEEE, 2018)
21. P. Boris, Predicting student drop-out rates using data mining techniques: a case study, in *Communications in Computer and Information Science* (2018), pp. 111–125
22. S.S. Ahmad Tarmizi, S. Mutalib, N.H. Abdul Hamid et al., A case study on student attrition prediction in higher education using data mining techniques. Commun. Comput. Inf. Sci. **1100**, 181–192 (2019). https://doi.org/10.1007/978-981-15-0399-3_15
23. K. Kang, S. Wang, Analyze and predict student dropout from online programs, in *ACM International Conference Proceeding Series* (2018), pp. 6–12. https://doi.org/10.1145/3193077.319 3090
24. W.F. Wan Yaacob, N. Mohd Sobri, S.A.M. Nasir et al., Predicting student drop-out in higher institution using data mining techniques. J. Phys. Conf. Ser. **1496** (2020). https://doi.org/10. 1088/1742-6596/1496/1/012005
25. E. Yukselturk, S. Ozekes, Y.K. Türel, Predicting dropout student: an application of data mining methods in an online education program. Eur. J. Open, Distance E-Learn. **17**, 118–133 (2018). https://doi.org/10.2478/eurodl-2014-0008
26. C. Pierrakeas, G. Koutsonikos, A. Lipitakis et al., The variability of the reasons for student dropout in distance learning and the prediction of dropout-prone students, in *Machine Learning Paradigms* (Springer International Publishing, 2020)
27. L. Kemper, G. Vorhoff, B.U. Wigger, Predicting student dropout: a machine learning approach. Eur. J. High. Educ. **10**, 28–47 (2020). https://doi.org/10.1080/21568235.2020.1718520
28. R. Da Fonseca Silveira, M. Holanda, M. De Carvalho Victorino, M. Ladeira, Educational data mining: analysis of drop out of engineering majors at the UnB – Brazil, in *Proceedings of the 18th International Conference on Machine Learning and Applications (ICMLA 2019)* (2019), pp. 259–262. https://doi.org/10.1109/ICMLA.2019.00048
29. D. Kim, S. Kim, Sustainable education: analyzing the determinants of university student dropout by nonlinear panel data models. Sustainability **10**, 1–19 (2018). https://doi.org/10. 3390/su10040954
30. N. Nurmalitasari, Z.A. Long, M. Faizuddin, M. Noor, Data preparation in predictive learning analytics (PLA) for student dropout. Int. J. Innov. Technol. Explor. Eng. **9**, 116–120 (2020). https://doi.org/10.35940/ijitee.c1027.0193s20
31. P.M. Da Silva, M.N.C.A. Lima, W.L. Soares et al., Ensemble regression models applied to dropout in higher education, in *Proceedings of the 2019 Brazilian Conference on Intelligent Systems (BRACIS 2019)* (2019), pp. 120–125. https://doi.org/10.1109/BRACIS.2019.00030
32. A. Mayra, D. Mauricio, Factors to predict dropout at the universities: a case of study in Ecuador, in *IEEE Global Engineering Education Conference (EDUCON 2018)*, April 2018, pp. 1238–1242. https://doi.org/10.1109/EDUCON.2018.8363371
33. T. Hasbun, A. Araya, J. Villalon, Extracurricular activities as dropout prediction factors in higher education using decision trees, in *Proceedings of the 16th International Conference on Advanced Learning Technologies (ICALT 2016)* (2016), pp. 242–244. https://doi.org/10.1109/ ICALT.2016.66
34. A. Saranya, J. Rajeswari, Enhanced prediction of student dropouts using fuzzy inference system and logistic regression. ICTACT J. Soft Comput. **06**, 1157–1162 (2016). https://doi.org/10. 21917/ijsc.2016.0161
35. F.J. da Costa, M. de S. Bispo, R. de C. de F. Pereira, Dropout and retention of undergraduate students in management: a study at a Brazilian Federal University. RAUSP Manag. J. **53**, 74–85 (2018). https://doi.org/10.1016/j.rauspm.2017.12.007
36. N. Tasnim, M.K. Paul, A.H.M.S. Sattar, Identification of drop out students using educational data mining, in *2nd International Conference on Electrical, Computer and Communication Engineering (ECCE 2019)* (2019), pp. 7–9. https://doi.org/10.1109/ECACE.2019.8679385

37. L. Wang, H. Wang, Learning behavior analysis and dropout rate prediction based on MOOCs data, in *Proceedings of the 10th International Conference on Information Technology in Medicine and Education (ITME 2019)* (2019), pp. 419–423. https://doi.org/10.1109/ITME. 2019.00100
38. A.A. Mubarak, H. Cao, W. Zhang, Prediction of students' early dropout based on their interaction logs in online learning environment. Interact. Learn. Environ., 1–20 (2020). https://doi.org/10.1080/10494820.2020.1727529
39. J. Liang, J. Yang, Y. Wu, L. Zheng, Big data application in education: dropout prediction in Edx MOOCs, in *Second International Conference on Multimedia Big Data* (IEEE, 2016), pp. 6–9
40. W. Li, M. Gao, H. Li et al., Dropout prediction in MOOCs using behavior features and multi-view semi-supervised learning, in *Proceedings of the International Joint Conference on Neural Networks 2016*, October 2016, pp. 3130–3137. https://doi.org/10.1109/IJCNN.2016.7727598
41. C. Burgos, M.L. Campanario, D. de la Peña et al., Data mining for modeling students' performance: a tutoring action plan to prevent academic dropout. Comput. Electr. Eng. **66**, 541–556 (2018). https://doi.org/10.1016/j.compeleceng.2017.03.005
42. K. Coussement, M. Phan, A. De Caigny et al., Predicting student dropout in subscription-based online learning environments: the beneficial impact of the logit leaf model. Decis. Support Syst. **135**, 113325 (2020). https://doi.org/10.1016/j.dss.2020.113325
43. B.R. Cuji Chacha, W.L. Gavilanes López, V.X. Vicente Guerrero, W.G. Villacis Villacis, Student dropout model based on logistic regression, in *Applied Technologies*. Communications in Computer and Information Science (CCIS), vol. 1194 (2020), pp. 321–333. https://doi.org/10.1007/978-3-030-42520-3_26
44. P. Strecht, J. Mendes-Moreira, C. Soares, Merging decision trees: a case study in predicting student performance, in *Advanced Data Mining and Applications*. Lecture Notes in Computer Science (Including Subseries Lecture Notes in Artificial Intelligence and Lecture Notes in Bioinformatics), vol. 8933 (2014), pp. 535–548. https://doi.org/10.1007/978-3-319-14717-8_42
45. W. Xing, D. Du, Dropout prediction in MOOCs: using deep learning for personalized intervention. J. Educ. Comput. Res. **57**, 1–24 (2018). https://doi.org/10.1177/0735633311875 7015
46. T.M. Barros, P.A.S. Neto, I. Silva, L.A. Guedes, Predictive models for imbalanced data: a school dropout perspective. Educ. Sci. **9** (2019). https://doi.org/10.3390/educsci9040275
47. A. Viloria, J.G. Padilla, C. Vargas-Mercado et al., Integration of data technology for analyzing university dropout. Procedia Comput. Sci. **155**, 569–574 (2019). https://doi.org/10.1016/j.procs.2019.08.079
48. I. Lykourentzou, I. Giannoukos, V. Nikolopoulos et al., Dropout prediction in e-learning courses through the combination of machine learning techniques. Comput. Educ. **53**, 950–965 (2009). https://doi.org/10.1016/j.compedu.2009.05.010
49. A. Behr, M. Giese, K.H.D. Teguim, K. Theune, Early prediction of university dropouts— a random forest approach. Jahrb. Nationaloekon. Stat. (2020). https://doi.org/10.1515/jbnst-2019-0006

Color-Based Plastic Bottle Cap Sorter Using ESP8266

Raja Fazliza Raja Suleiman, Ahmad Syahrul Najmi Zulkarnain, and Mohamad Maaroff Bahurdin

Abstract According to the plastic recycling industry, plastic material needs to be sorted by color to make the color of the plastic purer. Several color sorting machines exist in the industry, but the problem is that most of the machines are for large-scale industry, have huge dimensions and are expensive. Thus, this research work proposes to design and fabricate a low-cost portable bottle cap sorting machine based on color code using an ESP8266 and Arduino Uno. The machine operation can be monitored via an IoT (internet of things) platform called Blynk. In this work, the color sensor TCS 34725 detects 7 different colors of plastic bottle caps. The input data is processed and sorted by the microcontroller. In response to the controller, the rotary disk delivers the bottle cap to the rotary slope which segregates the bottle cap to the corresponding container based on the identified color. For machine performance analysis, a few tests were conducted. The sensor is tuned first, and all 7 colors can be detected without error. The time response test is executed to analyze the suitable sensing distance. A distance of 2 mm has the minimal average deviation color compared to those obtained for 5 and 7 mm with a time response measured of 0.02 min per cap. A real-time connection test between Blynk and the controller gives 2 s delay. The developed lightweight prototype may help small industries to perform color sorting tasks using minimal cost and space, along with remote monitoring via Blynk.

R. F. Raja Suleiman (✉)
Centre for Women Advancement and Leadership, Universiti Kuala Lumpur, 50250 Kuala Lumpur, Malaysia
e-mail: fazliza@unikl.edu.my

R. F. Raja Suleiman · A. S. N. Zulkarnain · M. M. Bahurdin
Industrial Automation Section, Universiti Kuala Lumpur Malaysia France Institute, Bangi, Selangor, Malaysia
e-mail: syahrulnajmi2797@gmail.com

M. M. Bahurdin
e-mail: maaroff@unikl.edu.my

R. F. Raja Suleiman · M. M. Bahurdin
UniKL Robotics and Industrial Automation Center, Universiti Kuala Lumpur Malaysia France Institute, Bangi, Selangor, Malaysia

© The Author(s), under exclusive license to Springer Nature Switzerland AG 2023
A. Ismail et al. (eds.), *Advances in Technology Transfer Through IoT and IT Solutions*,
SpringerBriefs in Applied Sciences and Technology,
https://doi.org/10.1007/978-3-031-25178-8_3

Keywords Color sorting machine · ESP8266 · Arduino · IoT · Blynk

1 Introduction

The sorting machine reduces the total time taken to complete a whole cycle in a manufacturing process. The manual sorting process implies several problems to the operation management [1]. Before the arrival of new technology, most Malaysian companies relied on the conventional method of sorting involving human work and was not cost-effective [2]. The automation and robotic industry have put much effort into introducing automated sorting systems due to high demand worldwide [3]. The quality and practicality of the system product is an important criterion of any current industry [4]. For companies that deal with many items that need to be sorted, the traditional method was no longer viable. There are a few types of sorting machines available in the industries [5], which mostly are created based on the size, weight, and material of the product to be sorted. Sorting machine with a robotic arm is used in large-scale industries (such as electronics and automotive), which is practical and flexible to pick and place heavy objects. A sorting machine is a very practical and economical process in which it can sort any product from one point to another point based on its category.

The sorting system is currently used in various industries such as the food sector, agriculture, and recycling. In the food industry, thousands of goods must be kept in good condition. Most sorting systems in this industry rely on visual inspection or product size for quality control [6, 7]. The author in [8] stated that, because of the system's ability to detect errors based on visual appearance, the visual sorting system is well suited for the food industry. Any food sorting system, such as the apple sorting system, will typically have a vision sensor, conveyor belt, separator, and classifier [9]. Meanwhile, in the agricultural sector [10], the color sorting machine enhances the crop sorting process. Theorists and engineers continue to develop systems and machines to meet today's requirements to improve the efficiency of operations in modern industry and improve large-scale sorting distribution to lower operating costs. In the recycling industry, specifically for plastic: the collected material is with mixed colors. It needs to be sorted by colors to obtain purer plastics. The sorted colored plastic is also more convenient than the mixed, to be used next for the secondary processing such as the plastic pelletizing process. This is the focus of this research work, to sort plastic bottle caps based on different colors. Several color sorting machines exist in the industry, the problem is that most of the machines are for large-scale industry, have huge dimensions and are expensive [11].

Some of the sorting machines are reported to be very complex and rigid, unable to achieve the application required in certain work stages [9, 10]. The existing sorting machine uses different sensing mechanisms such as inductive and capacitive sensors. Comparing with the other sorting machine that sort based on their size, weight, and material: the prototype sorting machine based on color code has its advantages [12].

Its size, weight, and material are not important anymore because it only sorts the product based on the color assigned to it.

2 Methodology

2.1 Process Flowchart

Figure 1 shows the process flowchart, and Fig. 2 depicts the block diagram of the presented work. Firstly, motor 1 moves from the initial position of 0–90° to bring the bottle cap to the sensor, then another 90° to the slider. The sensor determines the color of the bottle caps. TCS 34725 is the color sensor used in this presented work which offers a wide range of applications: including RGB LED. The signal from the sensor is subsequently processed and converted into a voltage value, with different colors corresponding to different voltage values. This value is sent to the Arduino board to begin the sorting process. Motor 2 will then rotate the slider to the container according to its color. Motor 2 moves at seven different angles which are 0, 25, 60, 120, 150, and 180° based on the color that has been detected. If there are no bottle caps detected, the servo motor will stop moving until the color sensor detects the bottle caps. In addition, NodeMCU will send monitoring parameters to a data cloud storage system, which may then be accessible using an IoT platform (Blynk). The Blynk app is compatible with both iOS and Android devices, and widely used for IoT projects [13, 14]. Next, if the bottle caps are entirely sorted to their colors, the system will notify the user via the Blynk application and displays the current sorted bottle caps and operations status on the Blynk monitoring dashboard (refer Sect. 3.4).

2.2 Prototype Design

The prototype is designed (refer to Fig. 3) using the SolidWorks software. Most of the parts are fabricated using a 3D printer because the filament for the 3D printer is more affordable than other materials and easier to assemble. The bottle caps are inserted into the delivery tube for the first loading process. After that, the color sensor detects. Next, servo motor 1 will rotate 180° to the slider and lastly, servo motor 2 will sort the bottle caps according to their colors. The control panel ensures safety and neatness, as well as to prevent wire tangling and accidents. A support bottle cap under the rotary disk ensures that the distance between the bottle caps and the sensor is in a good position for efficient color detection.

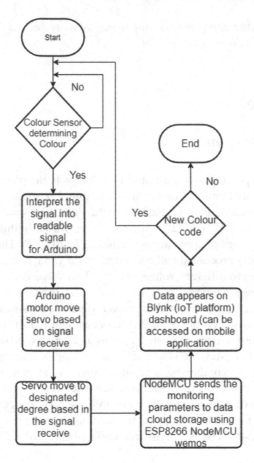

Fig. 1 Process flowchart

3 Results and Discussion

3.1 The Fabrication Prototype

Figure 4 shows the fabricated prototype. The model's base and structure are made of plywood with an aluminum frame. The container, LCD frame, and support servo motor are made of acrylic with 3 mm of thickness (transparency: easy to monitor the bottle caps). The 3D printing parts are designed using computer-aided design (CAD) software. Most of the 3D print parts are the most complex parts to fabricate (if using acrylic) which are the tube feeding holder, feeding tube, rotary disk, support servo motor, and slider. The length of this prototype is 600 × 600 × 341 mm, and the overall weight of the machine is 10 kg.

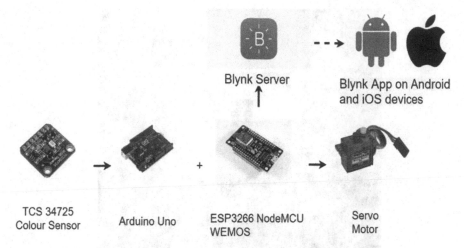

Blynk Server

Blynk App on Android and iOS devices

TCS 34725 Colour Sensor

Arduino Uno

+

ESP3266 NodeMCU WEMOS

Servo Motor

Fig. 2 Block diagram of the color-based sorting machine

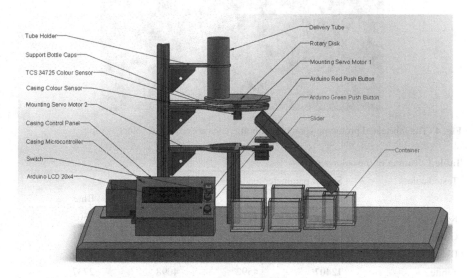

Fig. 3 Prototype design of the color-based sorting machine

3.2 Accuracy of the Color Sensor

The target of the test is to ensure that the whole system especially the color sensor is functioning properly. The color sensor TCS 34725 needs to be tuned (basic colors of red, green, and blue) to recognize the color of the bottle caps. Table 1 shows the color sensor tuning results to determine the 7 different colors. Tuning is conducted to maintain the efficiency of the color sensor and to get the best result possible without causing any confusion during the experiment.

Fig. 4 The fabricated prototype sorts plastic bottle caps according to colors

Table 1 Tuning of the color sensor

Tested colors	Sensor parameter reading			
	Clear	Red	Green	Blue
Red	3251	**2115**	632	502
Green	716	241	**251**	161
Blue	3969	910	1366	**1560**
White	**12407**	4592	4098	2737
Yellow	6000	**2624**	**2192**	831
Orange	3925	**2069**	**981**	567
Purple	3147	**1067**	867	**875**

The bold in the table corresponds to the matching of RGB sensor to the tested bottle cap color

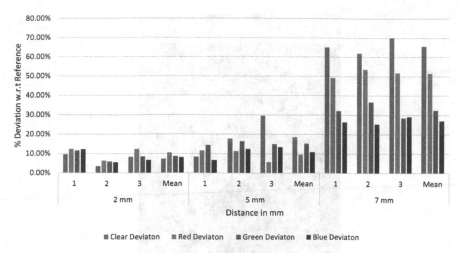

Fig. 5 Bar chart representing the sensing distance experiment

3.3 Placement of the Color Sensor

The experiment aims to figure out the best distance between the sensors and the bottle caps because the ambient lighting affects the sensor's detection quality. A specific distance is used in the experiment: 2, 5, and 7 mm. Each test is repeated 3 times. A specific distance can be determined by comparing the distance and effect of the distance mark with a specific distance and to achieve the best sensing distance. A red bottle cap is used as a reference for all other colors as shown in Table 1. Based on the result shown in Fig. 5, the most efficient distance between the color sensor and the object is 2 mm with constant color recognition. Referring to Table 1, the deviation concerning the reference value is calculated using Eq. (1) for 3 trials. The average deviation over the number of trials is calculated as well. For the tested distances, 2 mm has the minimal average deviation for the color clear (7.2%), green (8.7%), and blue (8.1%) compared to those obtained for 5 and 7 mm. Red deviation for 5 mm is the lowest (9.5%) compared to 2 mm, but the difference is nearly negligible.

$$\% \text{ Deviation} = \left(\frac{|x - \text{Reference}|}{\text{Reference}} \right) * 100 \tag{1}$$

3.4 Android-Based Graphical User Interface (Blynk)

Figure 6 shows the user interface of Blynk's application for this research work. When pressing the green push button at the mechanical part, the green LED at Blynk will

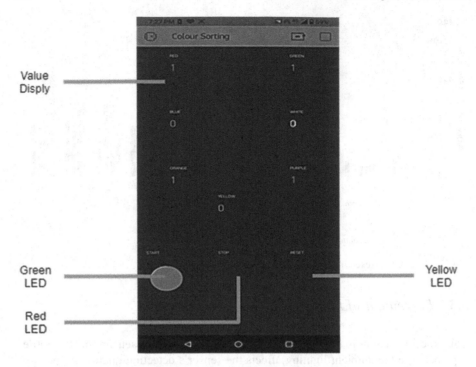

Fig. 6 Graphic user interface of the sorting machine for control and monitoring

light up to represent the system is starting. To stop the system, a red push button needs to be pressed and a red LED will turn on to indicate the system is stopped. Meanwhile, to reset the system to its initial condition press the yellow push button and a yellow LED will light up to show the system is reset. The current sorted color is also notified in the graphical interface.

3.5 Production Rate/Machine Performance

This sub-section presents the result of the system performance measurement which is sensor detection time and the ramp's movement speed. For the first measurement: three different colors were chosen as stated in Table 2. For each test, three measurements were taken using a stopwatch. The average measurement is calculated for each different color cap. Since this measurement was too fast, the measurement of four same-color caps was taken for verification purposes. The cycle time sensor detects 1 piece and 4 pieces of the same color red, green, and blue, as shown in the bar chart of Fig. 7. As a result, the average sensor detects red for 1.33 s, green for 1.26 s, and blue for 1.39 s for 1 piece. The average sensor detects 4 pieces of the same red color in 5.38 s, green in 5.30 s, and blue in 5.42 s. The average measurement of one

Table 2 Performance measurement of sensor's detection time

	Trial	Color detection time (s)		
		Red	Green	Blue
1 piece	1	1.67	1.43	1.76
	2	1.27	1.35	1.36
	3	1.05	1.01	1.05
Average		1.33	1.26	1.39
4 pieces (same color)	1	5.43	5.36	5.54
	2	5.32	5.30	5.35
	3	5.40	5.25	5.38
Average		5.38	5.30	5.42

cap and four caps are the same. The total average time taken to detect the cap color is about 1.34 s. The second measurement is the ramp's movement speed, which is timed using a stopwatch. Each experiment is repeated three times. The average of the ramp's movement speed is calculated using Eq. (2). Next, the cycle time (T_c) is calculated using Eq. (3). Lastly, the production rate per minute is calculated using Eq. (4). The final performances are shown in Tables 2 and 3. The cycle time for maximum ramp speed from 0 to 180° is shown in Fig. 8. The average cycle time for a ramp to travel from 0 to 180° at high speed is 1.21 s (0.02 min). The total cycle time, T_c, for one cap is equivalent to the sum of time to detect the cap color and the ramp change time. Thus, the total average time for one cycle is calculated using Eq. (5) that gives 0.04 min (2.4 s). The theoretical production rate (R_p) can be calculated using Eq. (4), yielding a production rate of 1500 caps sort per hour.

$$\text{Average measurement} = \frac{\text{Sum of measured values}}{\text{No. of trial}} \tag{2}$$

$$\text{Cycle Time } (T_c) = \frac{\text{Average of 1 piece}}{\text{Average of ramp maximum speed}} \tag{3}$$

$$\text{Production rate } (R_p) = \frac{60}{T_c} \tag{4}$$

$$\text{1 cycle average} = \text{Average of 1 piece} + \text{Average of ramp max speed} \tag{5}$$

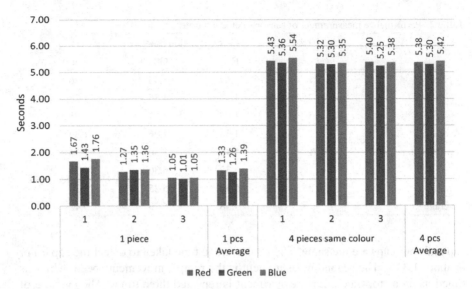

Fig. 7 Performance measurement of the sensor's detection by piece

Table 3 Performance measurement of ramp speed

Ramp speed	Trial	Cycle time (s)
Maximum ramp speed 0–180°	1	1.36
	2	1.27
	3	1.01
Average		1.21

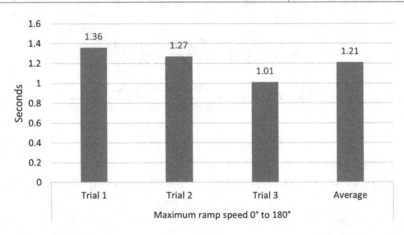

Fig. 8 Performance measurement of ramp speed by trial

4 Conclusion

The presented research work develops a bottle cap sorting machine based on color code with IoT ready. The machine sorts 7 colors and deploys a rotary disk with a servo motor. It is lightweight and inexpensive (10 kg and RM 386.50) compared to the usual available in the market. It may cater to small industries and allows remote control and monitoring from afar via the Blynk app, which can be used on smartphones and tablets. For future work, the hardware parts can be improved to smoothen the operating process such as bottle caps stuck in the tube due to the unalignment of the rotary.

References

1. A. Sachdeva, Development of industrial automatic multi color sorting and counting machine using Arduino nano microcontroller and TCS3200 color sensor. Int. J. Eng. Sci. **6**(4), 56–59 (2017)
2. N.N.S. Hlaing, H.M. Oo, T.T. Oo, Color detector and separator based on microcontroller. Int. J. Trend Sci. Res. Dev. **3**(5), 1103–1108 (2019)
3. A.K. Singh, M.S.R. Ali, Automatic sorting of object by their color and dimension with speed or process control of induction motor, in *2017 International Conference on Circuit, Power and Computing Technologies (ICCPCT)* (IEEE, 2017), pp. 1–6
4. R. Shah, A.B. Pandey, Concept for automated sorting robotic arm. Procedia Manuf. **20**, 400–405 (2018)
5. T. Henry, F. Jie, Design and construction of color sensor based optical sorting machine, in *2017 5th International Conference on Instrumentation, Control, and Automation (ICA)* (IEEE, 2017), pp. 36–40
6. R. Pourdarbani, H.R. Ghassemzadeh, H. Seyedarabi, F.Z. Nahandi, M.M. Vahed, Study on an automatic sorting system for Date fruits. J. Saudi Soc. Agric. Sci. **14**(1), 83–90 (2015)
7. M.M. Sofu, O. Er, M.C. Kayacan, B. Cetişli, Design of an automatic apple sorting system using machine vision. Comput. Electron. Agric. **127**, 395–405 (2016)
8. B.N. Mohammadi, H. Navid, J. Kafashan, Mechatronic components in apple sorting machines with computer vision. J. Food Meas. Charact. **12**(2), 1135–1155 (2018)
9. K.H. Wagh, D. Vilas, Automatic sorting using computer vision & image processing for improving apple quality. Int. J. Innov. Res. Dev. **4**(1), 11–14 (2015)
10. E.H. Yossy, J. Pranata, T. Wijaya, H. Hermawan, W. Budiharto, Mango fruit sortation system using neural network and computer vision. Procedia Comput. Sci. **116**, 596–603 (2017)
11. R. Tomari, M.F. Zakaria, A.A. Kadir, Z.W.N. Wan, W.M.H. Abd, Empirical framework of reverse vending machine (RVM) with material identification capability to improve recycling. Appl. Mech. Mater. **892**, 114–119 (2019)
12. L.B. Kachare, Object sorting robot using image processing. Int. J. Electron. Electr. Comput. Syst. **5**(7), 6–9 (2016)
13. R.F.R. Suleiman, F.Q.M.I. Reza, Gas station fuel storage tank monitoring system using internet of things. IJATSCE **8**(1.6), 531–535 (2019)
14. N.H.A. Rahim, A.N.H.M. Khatib, Development of PET bottle shredder reverse vending machine. IJATEE **8**(74), 24–33 (2021)

Cybersecurity for Online Safety Enhancement: Female Participation

Fatokun Faith Boluwatife, Zalizah Awang Long, Suraya Hamid, and Fatokun Johnson Oladele

Abstract Female participation in cybersecurity is low across the globe, as they represent below a quarter (about 11%) of global cybersecurity professionals. Cybersecurity being a crucial aspect of IT ought to involve women in the deliberation and proffering of security mechanisms and solutions; however, their representation is very minimal. Via a qualitative approach, this chapter discusses emphatically on the urgent need for female participation in cybersecurity. Involving females in cybersecurity has the potential of enhancing online safety, especially for children, as women naturally are concerned about the safety of their wards both online and offline. Thus, it is imperative to engage more females in the conceptualization, design, deliberation, development, up till the implementation of cybersecurity solutions as this can help to enhance online safety. Apparently, to achieve an effective participation of females in cybersecurity, it is reasonable to immerse cybersecurity knowledge alongside awareness into young female children early enough so as to break the gender disparity as well as establish a long-lasting interest of cybersecurity in females, thus contributing immensely to online safety enhancement. The review revealed that female IT employees' financial awareness can be impacted by investment decision relationships. Novel cybersecurity solutions should include special consideration for the female gender that can ensure that they are aware of female-targeted cyberattacks. Personal factors such as personality traits, abilities and interests, self-efficacy, gender

F. F. Boluwatife (✉) · Z. Awang Long
Malaysian Institute of Information Technology, Universiti Kuala Lumpur, Kuala Lumpur, Malaysia
e-mail: fatokun.faith@s.unikl.edu.my

Z. Awang Long
e-mail: zalizah@unikl.edu.my

S. Hamid
Faculty of Computer Science and Information Technology, Universiti Malaya, Kuala Lumpur, Malaysia
e-mail: suraya_hamid@um.edu.my

F. J. Oladele
Department of Mathematical Sciences, Faculty of Science, Anchor University, Lagos, Nigeria
e-mail: jfatokun@aul.edu.ng

identity, among others linked to cybersecurity, if negatively manifested, can result from exposure dearth to computing mentors and role models as well as mastery of experiential knowledge.

Keywords Cybersecurity · Women in IT · Women in cybersecurity · Female participation in cybersecurity · Women involvement in technology · Online safety

1 Introduction

There is a major gender imbalance in the technology industry, especially the area of cybersecurity. Periodically, women only make about 11% of the cybersecurity workforce globally [1, 2], a proportion considered biased. Female participation in cybersecurity has been relegated to dominance of cultural conceptions with regard to striking the balance between technology and gender [3]. This is applicable to women in the cybersecurity field as literature reveals their constraint by masculinity and security notions. Moreso, the societal hegemonic masculinity asserts that traits of the female gender are not linked with employment in security professions, as women are mostly associated to be more private, thence resulting in arguments that they ought to be protected instead of being agents of protection [4]. This assumption is among reasons why female participation in cybersecurity, alongside other IT fields is further reduced by the security facets of the cybersecurity profession.

Cybersecurity refers to the protection of information alongside hardware and systems that ensures the utilization, storage as well as transmission of information [5]. Moreover, cybersecurity involves series of activities tailored at protecting IT components and information, including the human from both online and offline attacks. It can also be addressed as the state of being secured from threat, putting into consideration the professionals that will ensure online activities are improved and secured. Commonly, this industry is not an heavy interest for women due to lack of awareness, and the mindset of being rejected as it is considered a male-dominant IT job [6].

Cybersecurity involves a complex process encompassing numerous benefits for users, such as enforcement of data privacy compliance for business by respective authorities. Moreover, cybersecurity is useful for almost all aspects of a nation, encompassing industry, companies, personal and the government systems. Cyber-security if mismanaged can pose a threat to national security [7]. There are several areas and specializations in cybersecurity, some of which includes cloud security, data loss prevention, application security, incidence response and forensic analysis, network defense, endpoint protection, penetration testing, internet of things (IoT) security, critical infrastructure security, secure DevOps, among others. Some of the job opportunities available in cybersecurity profession includes cloud engineer, technology projects lead, service engineers, cyber risk specialist, cloud administrator, cloud architect and cybersecurity consultant.

Literature reveals a gender disparity with regard to cybersecurity participation. Cybersecurity is being affected by gender disparity which poses discouragement to females, as they represent below a quarter (about 11%) of global cybersecurity professionals [5, 8]. Moreso, it has also been reported that only a very few attain leadership positions in the field of cybersecurity. This is especially disturbing as the male gender has dominated the fields of information technology (IT) [6], with cybersecurity being a major facet. Nevertheless, there is a necessity to boost female participation into various information technology fields, especially cybersecurity as they will be tremendously beneficial to the cybersecurity sector. However, this might not be an easy task but is achievable as the cybersecurity interest needs to be immersed into the female child right from young age probably via training in elementary and secondary schools.

Involving females in cybersecurity has the potential of enhancing online safety, especially for children, as women naturally are concerned about the safety of their wards both online and offline. Some vital lessons were gathered from an incidence tailored towards the special development of Facebook Messenger for Kids Application [9]. Here, a lot of privacy concerns and other issues were drawn by a coalition of organizations, led by the Campaign for a Commercial-Free Childhood. They issued an open letter to Facebook CEO, asking him to stop Facebook Messenger Kids due to their major concerns on the fact that the kids are too young and may not be able to manage security/privacy issues online. Interestingly, the bulk of members in this organization are women. This shows how much of attention and carefulness women put on security issues, thus the need to increase their participation in cybersecurity [10, 11].

The remainder part of this chapter discusses the methodology used in this review, results and discussions majoring on female participation in cybersecurity and drawing out critical reasons for urgent involvement of women in cybersecurity, and a conclusion on the overall appraisal of the topic. This chapter aims at contributing to advising and recommending the need for IT/Cybersecurity organizations to consider women in their cybersecurity policy development as well as software organizations in providing novel and impactful cybersecurity solutions that can help eradicate or mitigate the cybercrime rampant contemporarily.

2 Methodology

This chapter employs a qualitative research approach, involving critical review of literature on the participation of women in cybersecurity. It is more of a conceptual and narrative research, where case studies and empirical studies have been carefully reviewed to gather facts about the subject topic. Articles discussed were retrieved from high-impact journals, and research databases such as Science Direct, Google Scholar, IEEE, Springer, amongst others. The criteria for reviewing an article were based on the recency and relevance of the article to the current chapter. Articles more than 5 years as of the time of this chapter publication have been excluded from the review. The only articles above 5 years included are those termed as foundational articles. Also, studies with limited contribution were exempted as well. Data analysis was done via carefully transcribing and extracting the most important information needed to inform the purpose of this topic.

3 Results and Discussion

The major extractions from the review are presented as findings in this section. The chapter discussed in general about female participation in cybersecurity for online safety enhancement. First, it is important to discuss the women participation in IT generally, followed by their current participation in cybersecurity, and the need for more involvement of women in cybersecurity and how this can help enhance online safety.

3.1 Female Participation in IT

The issue of underrepresentation of women in the IT industry is not new, as it has been a concern right from time. In foundational research, the authors analyzed the state of women in IT occupations in the UK, it was revealed that the UK IT industry experienced segregation in gender occupation of which the onyx lies on females [12]. It is suggested that though women are not excluded from IT industry, they are likely to not attain higher positions and also may not be retained in the field due to gender disparity. One of the crucial features in the contemporary world is the emergence of information technology as it provides a platform for social, economic and developmental progress of countries across the globe, to be an efficient tool of which females are the major contributors. In a survey study that investigated how women-inclined entrepreneurship is impacted by information technology in Iran, it was found that IT with more female participation is of much significance in job access facilitation as well as enhancing product sales and marketing [13]. Thence, it can be arguably accepted that females play a vital role in the IT industry of which they are less represented and accepted. Literature has revealed that women are able to meet shortfalls of cybersecurity workforce development, develop innovative business strategies for the better cybersecurity assurances, as well as achieve most challenging concerns linked to the strategy of managing organizational technology [1, 6].

Furthermore, IT could link to other aspects of national building such as investments. Investment can be influenced by various factors such as conduct, investment decision attitudes, as well as financial knowledge. A recent study revealed that female IT employees' financial awareness can be impacted by investment decision relationships [14]. Surprisingly, the information technology sector is among sectors being sought for job by women. Female-related IT jobs include software developers, sales representative, project manager, IT administrator, customer service specialist, digital marketers, IT support/help desk, data analyst, financial analyst and graphic designers. However, there is lack of female professionals in the cybersecurity workforce.

Nevertheless, due to the work balance restraint in the IT industry, the majority of women feel demotivated [15]. There are huge consequences linked to the bias in the representation of females in the cybersecurity workforce, especially when putting the anticipated need for resource diversity and depth into consideration. As indicated in literature, diverse talent is a demanding skill for cybersecurity jobs. Nevertheless, employers should ensure that the cybersecurity industry is more welcoming and hospitable for females. Thus, the retainment and attraction of qualified workers, especially females, is a crucial issue in the field of cybersecurity which calls for an urgent attention [16]. This could be an area employers may want to work on to ensure more women feel safe and comfortable in the IT industry as they have much to offer in boosting the standard of the IT sector.

Interestingly, apart from just ensuring that females participate in IT, there are studies which also develop IT solutions that are designed with women in view. One of these studies is a work which surveyed about a security system for women and children [17], thus allowing them immediate response in case of harassment in public places in the society. The project was able to proffer dual solutions: one is the development of a portable device with a pressure switch. The pressure device has the ability of sensing pressure in case of an attack on a woman/child, after which the instant location of the user will be sent to the parent/guardian cell phone, as well as a call forwarded to the police for immediate rescue. Though there are very minimal IT solutions focusing on female gender, it is an eye-opener for IT solution providers to put the female gender into consideration while developing emerging IT solutions.

3.2 Gender Gap in Cybersecurity

The proliferation of cybercrimes of recent is a growing concern. There is a need for employers to source for security experts that can help curb the menace. At present, males are dominating major information technology fields, especially cybersecurity [6]. Moreover, cybersecurity is a field dominated by men across the globe. A study reported that women in North America only represented about 14% of overall cybersecurity workforce, whereas in Europe and Middle East, there are even lesser female cybersecurity professionals, 7% and 5% respectively [18]. The preceding statistics is rather alarming for a global cybersecurity industry. As estimated in literature, if there is a gender equality among cybersecurity professionals, this could lead to a rise in the cybersecurity industry economic footprint by 30.4 billion dollars in USA and 12.6 billion Pounds in the UK, respectively [5]. Nevertheless, there is a need to include more females to work as cybersecurity experts as this has the potential of tremendously benefiting any organization. As reported in literature, women represent only about 11% of cybersecurity professionals across the globe [1], see Fig. 1. This is a very low proportion, and such gender disparity is particularly discouraging for

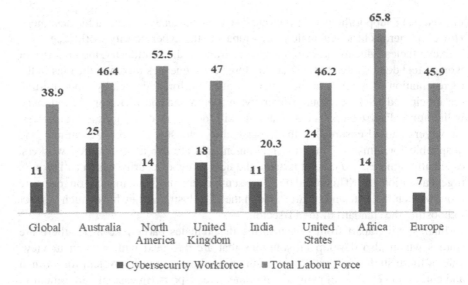

Fig. 1 Female representation in cybersecurity workforce and total labour force [5]

women. Moreso, only very few females who are already in the cybersecurity sector have the chance of rising to leadership positions.

Consequently, cybercrime is also exacerbated based on gender disparity. For example, more women had experienced stolen identities as compared to men. Women and girls are more likely to be targets of remote sexual abuse as compared to men counterparts [11]. Such cybercrime which coerces the female gender into posing online nudes as well as getting stalked online is also a major reason why females should be in the forefront of the fight against cybercrime as they would have more to offer in providing intelligent cybersecurity solutions that can counter salient aspects of cyberattacks which may have been relaxed by the male counterparts. Furthermore, security technologies are also limited for women and children, basically those referred to as weaker agents. For instance, there are issues regarding facial identification of women via the use of biometric facial recognition systems [1, 11]. Also, operators and security systems at airports have been known to unreasonably flag black women for certain unconfirmed offences as compared to the male counterparts. If the females are in charge of cybersecurity issues, they would be able to judge appropriately when carrying out cybercrime investigations [2].

The males predominate the role of cybersecurity professionals who protect software systems, databases and unauthorized access to computer networks. The future of cybersecurity is dependent on its ability to ensure women are attracted to the field, retained and promoted [6, 19]. This is essential as the female gender represents a resource that is under-tapped and highly skillful, especially in the field of information technology. It is also important for the cybersecurity field to put into consideration experiences of females as cybercrime victims as well as learn about steps that can help in addressing harm imbalances.

3.3 Increasing Female Participation in Cybersecurity

From the review conducted in this chapter, it is imperative to state that there is a large gap regarding the participation of females in cybersecurity. Overall, the female gender is a major player in issues of cybersecurity. Studies have suggested that females regardless of their background are potentials of novel ideologies, intuition and skills needed by every industry for achieving vigorous growth [3]. Corroboratively, a recent study [18] emphasized on the importance of women in cybersecurity as they offer unique perceptions. Explicitly, females usually have different perspective and approach of combating risks. They are typically risk-aversive, thus focusing more on compliancy alongside embracing organizational controls as compared to male counterparts [16]. Moreover, the female gender expresses higher level of intuitiveness and possesses a natural tendency towards social and emotional intelligence skills, a trait that puts them under calmness in the course of turbulence. This fact is essential for threat intelligence, incident handling, remediation efforts during a breach in security, as well as cybersecurity forensics.

Thence, it is only reasonable for organizations who aspire to enhance their cybersecurity measures to promote female participation in cybersecurity. Trainings need to be conducted specially for the females so as to ensure certain aspects of security are embedded that are more affective to the female gender. For example, many persons may not take the issue of certain cybercrimes such as online stalking, social engineering and love scams seriously, and unfortunately the victims of such crimes are females.

Moreover, novel cybersecurity solutions should include special consideration for the female gender that can ensure they are aware of female-targeted cyberattacks. Personal factors such as personality traits, abilities and interests, self-efficacy, gender identity, among others linked to cybersecurity [20], if negatively manifested, can result from exposure dearth to computing mentors and role models as well as mastery of experiential knowledge.

Linking female participation to success of cybersecurity missions in organizations, it is imperative to note that cybersecurity professionals can be better if they employ a holistic approach to bridge the gap between math, engineering and science to policymaking [10]. Also, to ensure advancement in cybersecurity, experts must possess a strong background in various technical fields as well as exert ability of transforming knowledge into law and orderliness, government policy and external regulations. Collaboratively, females are known to be more social and soft-skilled humans, thus, if they are part of cybersecurity professionals, they could offer more technical as well as interpersonal skills in battling cyberattacks more efficiently.

4 Conclusion

This chapter discusses emphatically on the urgent need for female participation in cybersecurity. Explicitly, females usually have different perspectives and approaches to combating risks. They are typically risk-aversive, thus focusing more on compliancy alongside embracing organizational controls as compared to male counterparts [16]. Due to the nature of the female gender encompassing their gentle and careful spirit, it is imperative to engage more females in the conceptualization, design, deliberation, development, up till the implementation of cybersecurity solutions as this has the potential of enhancing online safety. From the findings in literature, it is established that a large gap exists with regard to participation of females in cybersecurity. Cybercrime is exacerbated based on gender disparity. For example, more women had experienced stolen identities as compared to men. Women and girls are more likely to be targets of remote sexual abuse as compared to men counterparts [11]. The males predominate the role of cybersecurity professionals, thus the future of cybersecurity is dependent on its ability to ensure women are attracted to the field, retained and promoted [6, 19]. This is essential as the female gender represents a resource that is under-tapped and highly skillful, especially in the field of information technology. This cuts across employment, cybersecurity digital solutions, as well as victimization of the cybercrime itself. Females are prone to more vulnerability to cyberattacks, thus the need to ensure they are ably represented in the ecosystem of cybersecurity. Consequently, a non-biased gender talent acquisition, is considered to be the new capital for the contemporary world, thence empowerment of women in all professions, especially cybersecurity field, has the tendencies of propelling innovation as well as leverage untapped female talents for socioeconomic prosperity [6]. This study will help developers, stakeholders and organizations in setting the priority right with regard to women involvement as well as female participation in cybersecurity to ensure enhancement of online safety. Future studies should incorporate trainings as part of research studies to be conducted specially for the females so as to assess certain aspects of security which can be embedded into research as this can be effective to the female gender. Moreover, novel cybersecurity solutions should include special consideration for the female gender that can ensure they are aware of female-targeted cyberattacks. Personal factors such as personality traits, abilities and interests, self-efficacy, gender identity, among others linked to cybersecurity should be assessed in order to produce a comprehensive cybersecurity framework.

Acknowledgements The authors would like to acknowledge the support of Malaysian Institute of Information Technology, Universiti Kuala Lumpur for sponsoring this chapter.

References

1. N. Berríos, Increasing the participation of young women in cybersecurity. Comput. Sci. (2019). http://hdl.handle.net/20.500.12475/311
2. D.N. Burrell, Developing more women in managerial roles in information technology and cybersecurity, in *MWAIS 2019 Proceedings*, 20 (2019). https://aisel.aisnet.org/mwais2019/20
3. D. Peacock, A. Irons, Gender inequality in cybersecurity: exploring the gender gap in opportunities and progression. GST **9**, 25–44 (2017)
4. A. Withanaarachchi, N. Vithana, Female underrepresentation in the cybersecurity workforce—a study on cybersecurity professionals in Sri Lanka. Inf. Comput. Secur. **30**, 402–421 (2022)
5. N. Kshetri, M. Chhetri, Gender asymmetry in cybersecurity: socioeconomic causes and consequences. Computer **55**, 72–77 (2022)
6. R. Beveridge, Addressing the gender gap in the cybersecurity workforce. IJCRE **3**, 54–61 (2021)
7. Z.A. Zukarnain et al., Impact of training on cybersecurity awareness. GADINGST **3**, 114–120 (2020)
8. F.B. Fatokun et al., The impact of age, gender, and educational level on the cybersecurity behaviors of tertiary institution students: an empirical investigation on Malaysian universities. J. Phys. Conf. Ser. **1339**, 12–98 (2019)
9. D. Manson et al., The cybersecurity competition federation, in *Proceedings of the 2015 ACM SIGMIS Conference on Computers and People Research* (2015)
10. C. Willis-Ford, The perceived impact of barriers to retention on women in cybersecurity. Doctoral thesis, University of Fairfax, 2018. Accessed 17 July 2022
11. W.R. Poster, Cybersecurity needs women. (Nature Publishing Group, 2018)
12. A. Panteli et al., The status of women in the UK IT industry: an empirical study. Eur. J. Inf. Syst. **8**, 170–182 (1990)
13. L. Mivehchi, The role of information technology in women entrepreneurship (the case of e-retailing in Iran). Procedia Comput. Sci. **158**, 508–512 (2019)
14. C. Kathiravan et al., Women in the information technology sector's exploring financial knowledge and investment decision. PalArch's J. Archaeol. Egypt/Egyptol. **18**, 782–792 (2021)
15. T.B. Francis, P. Rajesh, Prevalence and patterns of work-life balance among women in the Information Technology Sector of Kerala, India. Int. Manag. Rev. **17**, 71–88 (2021)
16. S. Gonser, Jobs in cybersecurity are exploding: why are women locked out? (The Hechinger Report, 2018), https://hechingerreport.org/jobs-in-cybersecurity-are-exploding-why-are-women-locked-out/. Accessed 12 July 2022
17. S.K. Punjabi et al., Smart intelligent system for women and child security. IEEE IEMCON (2018). https://doi.org/10.1109/IEMCON.2018.8614929
18. H.G. Corneliussen, What brings women to cybersecurity? A qualitative study of women's pathways to cybersecurity in Norway. EICC (2020). https://doi.org/10.1145/3424954.3424965
19. L. Sharland, System update: towards a women, peace and cybersecurity agenda. (UNIDIR, Geneva, 2021). https://doi.org/10.37559/GEN/2021/03
20. T. Alharbi, A. Tassaddiq, Assessment of cybersecurity awareness among students of Majmaah university. Big Data Cogn. Comput. **5**, 23 (2021)

An Android Web Server-Based Real-Time Feeder Bus Tracking System

Zuhanis Mansor, Nurul Najwa Rusli,
and Siti Marwangi Mohamad Maharum

Abstract Although private transportation occasionally upgrades its engine specifications, design, usefulness, etc. the public transportation system has yet to receive the required improvements. Public transportation, particularly feeder buses, frequently fail to adhere to the timetable posted at station buses and is perpetually late for pickups. No application can view the current location of the bus, which is a concern for both train and bus passengers for public transportation. People these days want to be able to accomplish everything on a smartphone. In this paper, a bus tracking system that combines an Android application and a web server in real time is developed, enabling users to check the location of their desired transportation. The system can reduce waiting times and plan their travel or itinerary to avoid crowds, especially during peak hours.

Keywords Application programming interface · Global positioning system · Global system for mobile communications · Blynk application · Bus feeder

1 Introduction

Modern public transportation alternatives available in Kuala Lumpur and Selangor include the KTM commuter trains, the LRT light rail, the monorail, the RapidKL

Z. Mansor (✉) · N. N. Rusli
Advanced Telecommunication Technology Research Cluster, Communication Technology Section, Universiti Kuala Lumpur
British Malaysian Institute, Batu 8 Jalan Sungai Pusu, 53100 Gombak, Selangor, Malaysia
e-mail: zuhanis@unikl.edu.my

N. N. Rusli
e-mail: nurul.najwa@s.unikl.edu.my

S. M. Mohamad Maharum
Electronics Technology Section,
Universiti Kuala Lumpur British Malaysian Institute, Batu 8 Jalan Sungai Pusu, 53100 Gombak, Selangor, Malaysia
e-mail: sitimarwangi@unikl.edu.my

A. Ismail et al. (eds.), *Advances in Technology Transfer Through IoT and IT Solutions*,
SpringerBriefs in Applied Sciences and Technology,
https://doi.org/10.1007/978-3-031-25178-8_5

and BRT and the free bus service. The central hub and crossroads for all kinds of transportation is Kuala Lumpur. Currently, Kuala Lumpur and the surrounding Klang Valley area have various public transportation options, including buses, trains and taxis. However, due to its popularity, which is 1.79 million in the city and 6 million in its metropolitan area, Kuala Lumpur is experiencing the effects and challenges of rapid urbanisation and urban planning issues. To address the concerns of rapid urbanisation and urban planning, Kuala Lumpur (KLCH) intends to launch schemes that will improve the public transportation system and expand the Klang Valley's transportation management.

These countries' urban areas suffer from severe traffic congestion due to the rising number of cars on the road. One solution to these problems is using public transit such as LRT and MRT. In the Klang Valley, more than 1.20 million daily journeys or more than 20% of all trips in 2010 were made using the public transportation system [1, 2]. Hence, several things need to be fixed to raise the service level for buses, including their on-time performance.

The most popular modes of public transportation in Kuala Lumpur, Malaysia, are buses and trains. Most users are content to take buses or trains daily since they may avoid stopping issues, congestion during rush hours, affordable ticket rates and convenient services [3]. Additionally, it can reduce the use of private vehicles, petroleum use, and consequently, traffic congestion. However, no application can view the current location of the bus, which is a concern for public transportation. Public transportation, particularly feeder buses, frequently fail to adhere to the timetable posted at station buses and are perpetually late for pickups. Bus passengers are unsure of the bus's current location or how long they will have to wait for it to reach the stops (Fig. 1).

The GPS and ethernet-based real-time train tracking system was investigated by Rajkumar et al. in [4] to track the specific train accurately; the issues mentioned will be managed and taken control. It is not easy to follow the organisation buses, whereas moving on the highway. Sujatha et al. in [5] stated that phone discussion is not

Fig. 1 Missing bus passenger

continuously attainable due to traffic disturbances. In addition, the person boarding the bus might get irritated if they constantly get calls from the people boarding that bus. The authors of [6] investigated the satellite-based train monitoring system using the AVL VT300 GPS-GPRS module. The specific location, direction and other information about that train will always be apparent to the user.

A monitoring server and graphical user interface were implemented on the website using Microsoft SQL Server 2003 and ASP.net in [7] to enable them to view the vehicle's current location on a map on their website. The monitoring server establishes TCP/IP socket connections with the far-off hardware models. It is ready to speak with multiple client items, using a couple of threads. Their primary application add-ons are the socket conversation among the many monitoring servers, the web server and the GIS map server. Kalid and Rosli investigate the design of a system for school children's identification and transportation tracking in [8], in which the server and database will manage the backend operations, the KidBus. The tracker website is utilised with the support of mothers and fathers because the frontend interface instrument observes their kid's skills. The backend processing removes the noises from the data log and feeds the pieces of information into the website, allowing for visualisation of the raw data.

Based on the ideas and abovementioned limitations, this paper intends to simplify for users of trains and buses to use their smartphones to track their real-time location. This work can modernise Malaysia's existing transportation network and utilise geolocation to follow the movement in latitude and longitude coordinates. This study embarks on the following objectives.

i.   Design and develop the bus tracking system using two NodeMCU, a web server and a smartphone application.
ii. To enable the bus's location to be tracked in real-time using geolocation API.
iii. To save the user's waiting time by displaying the estimated time arrival of the bus using smartphone application.

2 Methodology

Figure 2 shows the block diagram of the prototype. It has three stages: the input unit, the controller unit and the output stage. There is a geolocation API in the input unit. NodeMCU is a component of the controller unit. The final one is the output unit, which is a web server and Blynk application. By form of power, the NodeMCU will be turned on. To enable the geolocation API and collect approximate location coordinates from public transportation, such as buses, the NodeMCU must first connect to a Wi-Fi network. The NodeMCU will then process this information. To calculate the distance and expected arrival time from the position to the bus station, the latitude and longitude are created in the NodeMCU. After verification, the data will be given to the web server and the Blynk application. In this work, the NodeMCU's role is to receive location data from the satellite, transform it into latitude and longitude positions, and then transfer that information to the Blynk

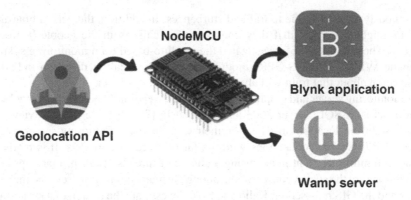

Fig. 2 Block diagram of the project prototype

application and web server in a format that the web server can understand. The Blynk application aims to inform the users on their smartphones of the estimated bus or public transportation arrival time and distance to the station. The web server will thus display the location of the bus.

Figure 3 demonstrates how the Blynk programme initiates when a bus passenger opens the Blynk app on their smartphone. The user must ensure their smartphone connects to the internet. Two tabs stand in for the bus services on the Blynk application's main screen. Next, the application will scan any Wi-Fi network nearby and connect to it and be able to the Blynk application connected to the NodeMCU and then request the location data of the transports (NodeMCU). The NodeMCU can determine the real-time location and predict the user's transport arrival time. If the Blynk application does not connect to any Wi-Fi network, the process returns, where the application will scan any Wi-Fi network nearby and connect to it. After the information of the location of the transport (NodeMCU), distance to the station (place) and the arrival time of the bus is recognised, the Blynk application displays the information. Then the user can exit the application.

According to Fig. 4, the project system's flowchart begins when the NodeMCU is powered on and searches for a connection to any available Wi-Fi network as it can connect to the geolocation API. Since the API provides a readable address, there is no need to transform the output format to another format. NodeMCU will be connecting to the geolocation API every 8 s. This will give a new real-time location of transport in position as latitude and longitude. If the data cannot be read, the step will return to where the original data was sent. If the NodeMCU is turned off, the whole process will end. Otherwise, the project will continue to connect to the Wi-Fi network. The process involves sending the location, distance and estimated arrival time (ETA) data to Blynk and the web server.

Fig. 3 The flowchart of the
Blynk application

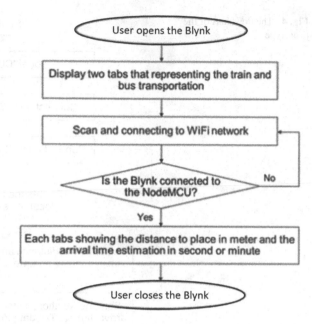

3 Results and Discussion

The development of the train and bus tracking system in Kuala Lumpur using a web server and an Android application resulted in a well-built end product, as depicted in Fig. 5. The NodeMCU microcontroller serves as the system's central nervous system. Additionally, it serves as an interface between the geolocation API, the Blynk application, and the web server, receiving, managing, and processing data input and output between the devices. The NodeMCU processed the latitude and longitude coordinates after receiving all the data from the geolocation API. This operation is instructed to update itself constantly every eight seconds.

The Blynk programme provided information about the distance and projected arrival time between the prototype and the target location, which is used to get the project's final output result. Users can also observe the transit location in satellite imagery with a web server. The findings indicate that the user must open the Blynk application and wait for the information to be displayed in order to track the anticipated distance travelled and arrival time of the bus and train. The user must make sure that their device is connected to the Blynk application. Figure 6 provides an example of this, showing the user at the right-wing female hostel and the target location as the left-wing female hostel, guard house and grill station. The arrival time is predicted in seconds or minutes, while the distance is displayed in metres. The web server is a form of website that is used to display the location of the bus and train (tracker) in mapping to the user. In order to see the website, users just have to browse and wait for website to load. This can be illustrated as in Fig. 7 where the user waits at the right wing of the hostel. Moreover, Fig. 8 shows the distance and arrival times for

Fig. 4 The flowchart of the
prototype

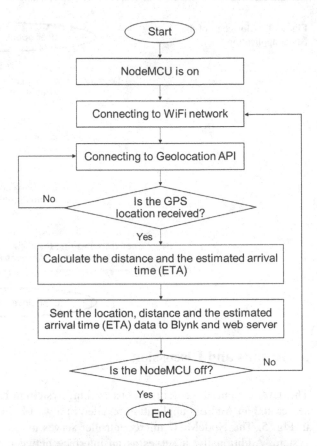

Fig. 5 Prototype of the
transport tracker

Fig. 6 The Blynk
application displays the
information

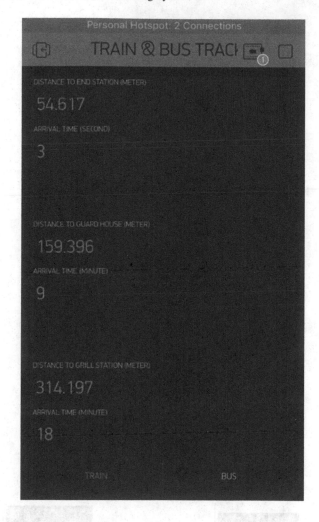

each point along the route as the tracker moves towards the end station (left wing of female hostel).

4 Conclusion

Malaysia's public transport could strengthen the great lifestyles and make life easier. It is the competitively priced means of travelling in particular for the pupil to access education. Nonetheless, the bus and train transport has a very terrible transportation knowledge system because it is founded on the timetable. In this paper, a fast-tracking real-time bus tracking system that includes an Android application and a web server

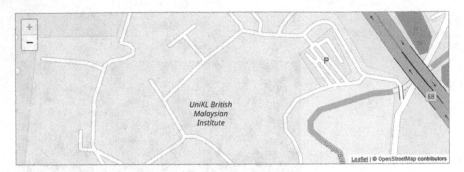

Fig. 7 The web server displays a map of the tracker's location

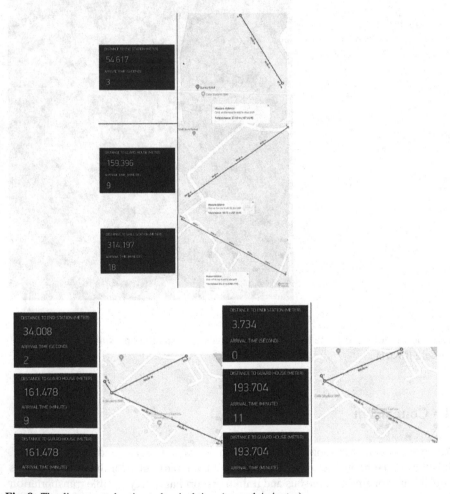

Fig. 8 The distance and estimated arrival time (seconds/minutes)

allows users to monitor the whereabouts of their preferred mode of transportation. During peak hours, the system can shorten wait times and organise their journey or schedule to avoid crowds. The ready time of the user can be lowered by using the IoT to inform the passenger via Wi-Fi. It resulted in the current bus and train system where it can be superior to be extra mighty and efficient so that users can create better planning.

Acknowledgements Zuhanis Mansor would like to thank the Advanced Telecommunication Technology (ATT) Research Cluster and Universiti Kuala Lumpur British Malaysia Institute (UniKL BMI) for the provision of laboratory facilities and financial support.

References

1. C.O. Chuen, R.M. Karim et al., Mode choice between private and public transport in Klang Valley. Sci. World J. **2014**, 14 (2014)
2. J.D.L. Margaret, Spatial dynamics of tour bus transport within urban destinations. Tour. Manag. **64**, 129–141 (2018)
3. Star Media Group Berhad, SPAD poll: more using public transport in KL now. (The Start Online, 2017), https://www.thestar.com.my/news/nation/2017/05/16/spad-poll-more-using-public-transport-in-kl-now/. Accessed 8 Oct 2018
4. R.I. Rajkumar, P.E. Sankaranarayanan et al., GPS and Ethernet based real time train tracking system, in *IEEE International Conference on Advanced Electronic Systems* (2013), pp. 282–286
5. K. Sujatha, P.V.N. Rao et al., Design and development of android mobile based bus tracking system, in *1st International Conference on Networks & Soft Computing* (2014), pp. 231–235
6. N. Das, C. Das et al., Satellite based train monitoring system. J. Electr. Eng. **36**(2), 35–38 (2011)
7. W. El-Medany, A. Al-Omary et al., A cost effective real-time tracking system prototype using integrated GPS/GPRS module, in *IEEE 6th International Conference on Wireless and Mobile Communications* (2018), pp. 521–525
8. K.S. Kalid, N. Rosli, The design of a schoolchildren identification and transportation tracking system, in *International Conference on Research and Innovation in Information Systems* (2017), pp. 1–6

Smart Parcel Box: Sanitizer and Security

Ainor Khaliah Mohd Isa and Nurul Shahira Azuddin

Abstract The COVID-19 pandemic and movement control order that started in 2020 has changed the shopping behavior to online shopping. This also increases in-home delivery services by the shipping providers. However, since the virus can be transmitted through surface transmission, the buyer is advised to avoid touching surfaces and clean or sanitize surfaces regularly with standard disinfectants to prevent the spread. In addition, with an increase in the parcel delivery process, missing parcels also will be one of the main problems that the buyer will be facing. This study has developed a smart parcel box with sanitizer to overcome the issues stated above. Smart in this context refers to the notification that the buyer will get once the parcel is placed inside the parcel box, the box itself will be locked once the parcel is in and only can be opened by the authorized user. This study utilizes the Arduino IDE software to control the operation of the locks and notifications. The notifications are linked to the Blynk Application that needs to be installed on the buyer's smartphone. The software is also coded to run the motor that controls the standard disinfectants that will be sprayed on the parcel for a few seconds. The results indicate one solution for the online shopping addicts to shop while avoiding the spread of COVID-19 viruses.

Keywords Online shopping · Covid-19 · Smart parcel box · Sanitizer · Blynk

A. K. Mohd Isa (✉) · N. S. Azuddin
Universiti Kuala Lumpur British Malaysian Institute, Batu 8, Jalan Sungai Pusu, 53100 Gombak, Selangor, Malaysia
e-mail: ainorkhaliah@unikl.edu.my

N. S. Azuddin
e-mail: nshahira.azuddin@s.unikl.edu.my

A. Ismail et al. (eds.), *Advances in Technology Transfer Through IoT and IT Solutions*,
SpringerBriefs in Applied Sciences and Technology,
https://doi.org/10.1007/978-3-031-25178-8_6

1 Introduction

The movement control order (MCO) in Malaysia that was implemented on March 18, 2022 has transformed the face of e-commerce. With most physical stores brought to a halt, travel restrictions and avoiding crowded places and long queues and precautionary measures have made the surge in online purchasing. Several studies [1, 2] to determine consumer buying behavior and factors that influence online shopping have been done in this pandemic era.

According to a recent study by the University of Guelph, when physical distance guidelines and proper cleaning techniques are followed, high-touch surfaces in grocery stores are less likely to be exposed to COVID-19. Furthermore, the issue of the missing packages is also quickly heard among customers. Based on the number of package thefts in the United States is increasing. In 2019 alone, 90,000 packages were lost every day in New York City, an increase of 20% from four years ago. Every day, more than 1.7 million pieces of goods are stolen or lost across the country, resulting in a loss of US$25 million in products and services [3].

Nowadays, most people in the world are busy with their life, so the smart parcel is needed among them to make their work easier. Next, the product which contains a technology system is a very high demand nowadays. Physical items that have been developed by people and produced via methods that work to serve specific functions are referred to as technological products. Technology connects people's abilities, knowledge, processes, techniques, and equipment to solve issues and make their lives safer and easier. Technology is vital today because it is pushing the world forward and is improving the world.

There was little research related to this project and has been the benchmark for developing this prototype. The comparison study in terms of the input, processing, and output has been accomplished.

In [4] the Smart Parcel Box with UV-based sanitization was developed, where the UV lamp was a medium to sanitize the package. The NodeMCU was a controller to control the system, and ESP32-CAM was a camera for the customer to monitor from far away. The second researcher [5] developed the internet of things (IoT)-based parcel monitoring system. In this project, the simple mail transfer protocol (SMTP) sends a notification to the customer to notify them about their packages. GPS is used for real-time location lock. Raspberry Pi is a controller to control the system. Twillio SMS API is used to send and receive SMS messages, track the delivery of sent messages and retrieve and modify the message history of the parcel. It was a complicated hardware and software design and need close monitoring. In [6], the Raspberry Pi is the microcontroller to control the system, relay to open and close the circuit, EM lock to lock the box, HTTP to store users' data, and unique box IDs and user IDs for each box to secure the box from unauthorized persons were used. Besides, PHP keeps the data in the Wamp server. Next, a barcode scanner application scans the barcode that appears on the raspberry pi and is also known as the product barcode. This barcode will be scanned by the courier to send the notification to the customer as their package has been delivered.

This paper presents the results of the development of a smart parcel box with a sanitizer prototype. The objective of this project was to design the appropriate hardware and software materials used to develop a smart parcel box that will also be operated with a sanitizer. The remainder of this paper is organized as follows: Sect. 2 provides a thorough description of the methodology comprising a block diagram and flowcharts. Section 3 presents results including the prototype and the monitoring applications. Finally, Sect. 4 summarizes the key findings of this work and provides directions for future work.

2 Methodology

This project is divided into two phases, i.e., software and hardware development. The first step was to produce a block diagram showing the input, processing element, and output. Figure 1 shows the block diagram of the project.

2.1 Block Diagram

The input for this prototype is a magnetic sensor located at the door. The sensor is connected to the heart of the system, NODEMCU ESP32. ESP32 is a range of low-cost, low-power system-on-a-chip microcontrollers that also include built-in Wi-Fi and dual-mode Bluetooth. This Bluetooth feature is the only content in the ESP32. The ESP32 is a low-power processor designed for mobile devices, wearable electronics, and IoT applications. It uses a combination of proprietary software to achieve ultra-low-power consumption. In addition, ESP32 has cutting-edge capabilities including great clock gating, several consumption modes, and dynamic power scaling. With an operating temperature range of −40 to +125 °C, the ESP32 can perform effectively in industrial conditions. Through its SPI/SDIO or I2C/UART interfaces, the ESP32 can communicate with other systems to provide Wi-Fi and

Fig. 1 Block diagram

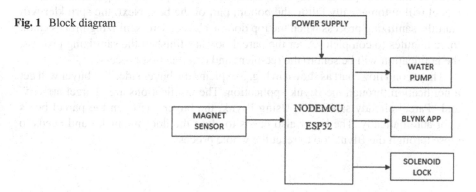

Bluetooth capability [7]. Then, the ESP32 is a microcontroller board, and it can be programmed using the Blynk app. In general, this microcontroller gadget comes with a Wi-Fi module that allows it to connect to an IoT system and save data in the cloud. As illustrated in the diagram, this microcontroller device will act as a device controller or process for the system, receiving input signals from both sensors and sending the signal or data to the output component for operation and display.

Based on the block diagram, there will be three outputs that are controlled by the microcontroller. Blynk application was connected to the microcontroller using Bluetooth. Blynk will show the data of the frequency of the magnetic sensor in terms of the date and time the door opens and closes. Blynk has also been programmed to send notifications to the buyer when there is a parcel in the box. The below door opening is also being controlled by the Blynk application. Blynk is a platform that allows users to create interfaces from either iOS or Android devices for controlling and monitoring user hardware projects. Blynk allows users to develop mobile applications that can easily connect to use an Arduino as easily as dragging a widget and setting up a pin [8].

The water pump is also being connected to the microcontroller. The water pump is used to pump up the sanitizer and is also connected to the water sprinkler. The water sprinkler was located at the below part of the box against the back wall of the box. The water sprinkler will be activated as soon as the door is closed, and the parcel is dropped off. The third output is the solenoid lock that is positioned on the below door.

2.2   Flowchart

The second stage is to produce a flowchart. In this project, two flowcharts were prepared to show the process of the parcel drop-off and user notification.

The first flowchart in Fig. 2 illustrates the drop-off process by the delivery man until the sanitizing process. The delivery man needs to open and close the top door manually to drop off the parcel. There is a flap that separated the top and bottom areas of this smart parcel box. Once the delivery man placed the parcel inside, the parcel will automatically fall to the bottom part of the box. Next, the sprinkler will start the sanitizing process when the top door is closed. This sanitizing process took three minutes to complete. After the parcel box has finished the sanitizing process, the notification will be sent to the recipient and end this first process.

The second flowchart as shown in Fig. 3 explains the buyer side. The buyer will get a notification through the Blynk application. The notifications are "Parcel arrived" and "Parcel already sanitized". Using Blynk, the buyer can open the parcel box's door automatically. The buyer also needs to close the door manually and needs to close through the Blynk too to reset the whole process.

Fig. 2 Flowchart 1

Fig. 3 Flowchart 2

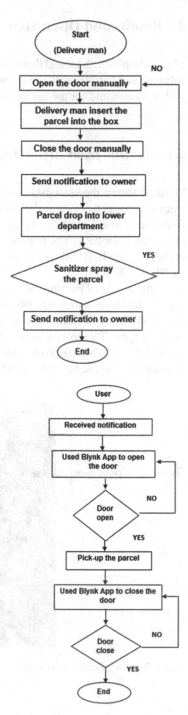

3 Results and Discussion

This section will show and discuss the results of the prototype. It is categorized into software and hardware. In the software part, screenshots from the Blynk application will be shown and for the hardware, the final prototype will be featured.

3.1 Results from Blynk App

Figure 4 shows the screenshot from the Blynk Application. The data collected are from the magnetic door at the front of the prototype. The time taken accumulated from the moment the delivery man opens the door to drop off the parcel until he closed the door back. The feature in this application was set through a set of coding in the Arduino IDE to show the date and time. This Blynk application was connected by Bluetooth.

Notifications of the presence of the parcel and sanitizing process have also been sent to the buyer using this Blynk application. The first notifications as shown in Fig. 5 will be delivered to the user as soon as the delivery man closed the door of the prototype. The second notification will be received once the sanitizing process is done. This sanitizing process took about 3 s to complete. Figure 5 also shows the notification that the buyer received.

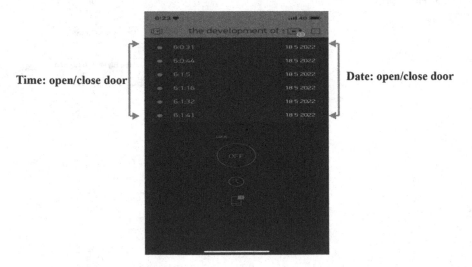

Fig. 4 Blynk application screenshot

Fig. 5 Notifications in Blynk

3.2 Prototype

The prototype of the smart parcel box used the existing parcel box that can be bought online and upgraded into a smart parcel box. Figures 6 and 7 show the front and rear views of the smart parcel box. From the front, there are a drop-off door and a collection door. The delivery man can only open it once to drop the parcel off. The collection door can be locked by the buyer. There is a solenoid lock inside the box that has been controlled by the NodeMCU ESP32 and the notification of the date and time of opening were sent to the buyer using the Blynk application. The rear of the smart parcel box is the most important location. This is the control panel of the prototype that consists of the NodeMCU ESP32, relay, and sanitizer pump. In a real application, the rear of the prototype cannot be seen by the public because it will be fixed onto the wall of the parcel box compartment.

4 Conclusion

The smart parcel box with sanitizer has been successfully developed. The initial stage of this project was to have a sketch of the overall ideas of the parcel box prototype that have brought to the development of the software (coding using Arduino IDE and Blynk application) and hardware (sensors and motor sprinkler) of the prototype. This resulted in the construction of the prototype structure (hardware) and software integration. The amount of sanitizer sprinkled is sufficient to cover the overall surface of a standard parcel. The data are shown in the Blynk application including the frequency of opening and closing the parcel box that can be further used in the future to monitor the rate of receiving parcels to enhance savings.

Fig. 6 Smart parcel box with sanitizer prototype

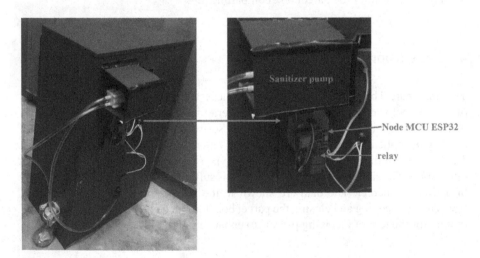

Fig. 7 Rear view of the parcel box, showing the control panel

Acknowledgements The authors would like to thank Universiti Kuala Lumpur, British Malaysian Institute (UniKL BMI), for providing a conducive environment and technical support for this project.

References

1. R. Musa, S. Nasaratnam, K. Rethinam, P.M. Varatharajoo, A. Shanmugam, A study of factors influenced online shopping behavior in Malaysia: a structural approach (2022). https://doi.org/10.37394/23207.2022.19.48
2. N.S. Mohamad Shariff, N.H.I. Abd Hamid, Consumers' buying behavior towards online shopping during the COVID-19 pandemic: an empirical study in Malaysia (2021). https://doi.org/10.33102/mjosht.v7i2.164
3. K. Schoolov, With package theft at an all-time high, Amazon and others are fighting back. (CNBC, 2020, Jan 11), https://www.cnbc.com/2020/01/10/package-theft-how-amazon-google-others-are-fighting-porch-pirates.html. Accessed 19 July 2022
4. K. Rohan, J. Kavish, L. Kiran, S. Krushankan, P. Madhura, Smart parcel box with UV based sanitization. (IRJET, 2020, Oct 13), https://www.academia.edu/44286447/IRJET_Smart_Parcel_Box_with_UV_based_Sanitization. Accessed 20 July 2022
5. T. Mallick, A. Kalam, N. Naveed, Internet of Things (IoT) based parcel monitoring system (2018)
6. Delivery box IoT (2018, May 7), https://harshaiot.wordpress.com/delivery-box-2/. Accessed 20 July 2022
7. NODEMCU ESP32 Cytron Technologies Malaysia (2022), https://my.cytron.io/p-nodemcu-esp32. Accessed 20 July 2022
8. Blynk IoT platform: for businesses and developers (2022), https://blynk.io/. Accessed 20 July 2022

Development of a Smart Trash Can/Dustbin Using Internet of Things

Mohd Zain Ismail, Siti Zaiummi Mohammad Zanawi, and Mohd Ridhuan Yusof

Abstract Malaysia has one major issue that is hard to avoid: overflowing garbage bins. This problem occurs due to a lack of monitoring from waste management authorities and public awareness. Because of that, a viable solution must be made. This study aims to tackle the problem by implementing technologies and Internet of Things (IoT) in the garbage bin system. The objective of this study is to determine the content of the garbage bin by using an ultrasonic sensor connected to Raspberry Pi. The use of Raspberry Pi has many advantages, such as a built-in database that can store the data, and act as a monitoring system, and a dashboard to show real-time data. The results obtained from this project are that the sensor successfully predicted the content of the garbage bin. When the bin reaches 80% capacity, the results obtained are sent to the MySQL database and presented to the Node-Red dashboard for further monitoring. Furthermore, an email notification was sent to waste management authorities to inform on the status of the garbage bin. In conclusion, this system has eased the waste management authorities' work by informing them when the garbage bin is full and also improving the cleanliness of the area.

Keywords Raspberry Pi · Node-Red · Garbage bin · MySQL · Ultrasonic sensor

M. Z. Ismail (✉) · S. Z. Mohammad Zanawi · M. R. Yusof
Communication Technology Section, Universiti Kuala Lumpur British Malaysian Institute, Gombak, Malaysia
e-mail: mzain@unikl.edu.my

S. Z. Mohammad Zanawi
e-mail: sitizaiummi@unikl.edu.my

M. R. Yusof
e-mail: mridhuan.yusof@s.unikl.edu.my

1 Introduction

Internet of Things (IoT) describes a network of real-world objects that have been equipped with sensors,software, and other technologies to enable Internet-based data transmission and communication. These gadgets include everything from common household items to complex industrial machines. IoT has emerged as one of the most important technologies of the twenty-first century in recent years [1].

Furthermore, data storage is simply one aspect of the IoT ecosystem, including data collecting, transmission, storage, computing, analyzing, and applications; nonetheless, it is still a very complex technology. Data integrity, dependability, and security are all requirements for IoT storage solutions. Additionally, the storage solution must be compatible with a variety of end devices, edge gateway, and data center environments [2].

Next, garbage bins are widely used around the world. They come in varieties of sizes: small, medium, and big.Although they come in many sizes, the problem with garbage bins that it is easy to be overflown with trash. This led to a lack of public awareness and ignorance the fact that the garbage bin was already full. If their trash has been rid of, they do not care about the condition of the dustbin itself. Not only that, the waste management authority that handles emptying the garbage bin only cleans it once a week without knowing on what day the garbage bin is full [3–7].

The objective of this study is to construct a system that can determine the amount of trash inside the garbage bin. When the garbage bin is almost full, 80%, the system sends out a notification to the authorities to empty the garbage bin as soon as possible. If ignored, the garbage bin can be overflown. The results obtained from the sensors are sent out to the MySQL database and displayed in real time using the Internet of Things platform. The IoT platform that is used in this study is the Node-Red dashboard.

Before beginning this study, fully detailed research must be made beforehand. This paper discovered previous research that is similar to this study. First, the paper is done by [8], designing a smart green environment of garbage monitoring system measuring the garbage level in real time and alerting the municipality whenever the bin is full based on the types of garbage. This project used an ultrasonic sensor to measure the garbage level, and an ARM microcontroller to control the system. The system used in these studies shows different statuses for different types of garbage such as domestic waste, paper, glass, and plastic [8].

A study made by [9] provided a solution to cleaning up the roads and streets in India, a garbage monitoring system using Raspberry Pi and GPS link. The system proposed to monitor the garbage level by attaching a sensor inside the garbage bin and providing the information to the municipality office. The components used in this project are Raspberry Pi, GPS, GSM, and IR sensors. The garbage level is sensed by IR sensors, and GSM is used to send messages to the user by an Android device's smartphone. The message sends the status and location of the garbage bin [9].

To support the green cities initiative, a project is being introduced. A study conducted by [10] planned and constructed an IoT E-waste management monitoring system that will provide a solution to electronics waste collecting and generating data. The system monitors the waste level and temperature and sends the information to the system administrator. An ultrasonic sensor, a flame sensor, a Raspberry Pi 3, and ThingSpeak as an IoT web platform were used. By proactively monitoring and controlling the collection of e-waste through the Internet of Things concept, this system is expected to increase the use of e-waste recycle bins, thus helping green city initiatives and producing a greener environment [10].

The information can provide improved ideas for completing this project based on earlier studies. The project will provide an IoT web platform for real-time data monitoring and a sensor for detecting the amount of trash in the rubbish bin.

2 Material and Methodology

The block diagram in Fig. 1 is divided into 3 parts: input, process, and output. At input, two ultrasonic sensors detect the amount of trash inside the garbage bin. When the sensors detect it, it sends the data to the Raspberry Pi to process. The Raspberry Pi must be linked to the Internet for the ultrasonic sensor to send data successfully; for this project, a wireless connection from the router and mobile hotspot is needed. After that, Raspberry Pi will process the data from the input and sends it to the MySQL database. Then, the database sends the complete results to the output. The first output is the email which sends a message to the waste management authorities about the status of the garbage bin. For output 2, the Node-Red dashboard displays the full results of the data obtained from the database in real time, meaning the data will refresh automatically.

The flowchart in Fig. 2 shows the start of the system. When Raspberry Pi is turned on and connected to the Internet via a mobile hotspot, the ultrasonic sensor will read the level of the garbage bin and determine the content. The sensor detects three statuses for the garbage bin: empty, half-full, and full. The empty status is when the garbage bin has been empty and little trash accumulated inside it. Half-full is when the trash occupies half of the bin and full status is when the trash has almost accumulated the entire bin. When the sensor detects the status of the garbage bin, the system sends out an email notification to the authorities. Furthermore, the full data that consists of the date, time, content, and status of the garbage bin is sent to the MySQL database for storing and monitoring the data. From the authority's side, they can monitor the data in real-time using the Node-Red dashboard that has been provided.

Figure 3 shows the prototype sketch that represents what the system would look like. The sensors are placed at the top of the lid for the sensors to easily read the content of the garbage bin. Raspberry Pi and the power supply which is the power bank are placed at the back of the garbage bin.

This project's code is made by using Node-Red, and its programming language is based on JavaScript. Node-Red has simplified the process by providing a browser-based flow editor for programming as shown in Fig. 4. The browser interface is

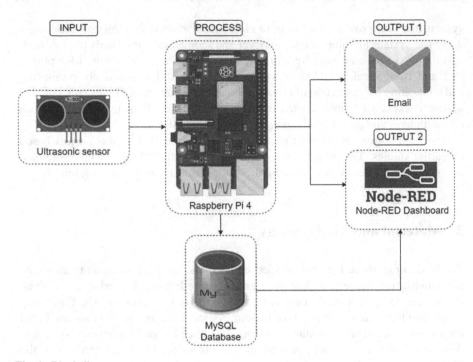

Fig. 1 Block diagram

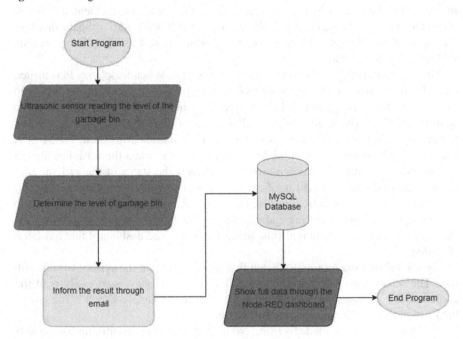

Fig. 2 Flowchart of the project

Fig. 3 Project sketch

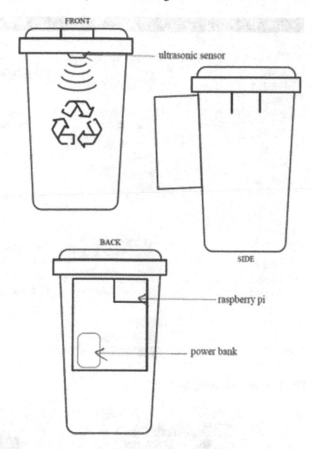

divided into three sections: palette, workspace, and sidebar. The palette has numerous nodes that can be used in this project. Each flow is built in the workspace. The sidebar displays node information, debug messages, configuration nodes, and context data for each node.

3 Results and Discussion

The smart dustbin is shown in Fig. 5. The system operates on Raspberry Pi 4 that is powered by a power bank. A mobile network connection is used to connect Raspberry Pi. Without an Internet connection, the system cannot send the data to the output. Figure 5 shows the project's complete prototype.

Figure 6 shows that the ultrasonic sensor is placed on the top of the lid. When the lid is closed, the sensor reads the distance inside the dustbin. The sensor reads the distance in centimetres value and converts it to a percentage for easier understanding.

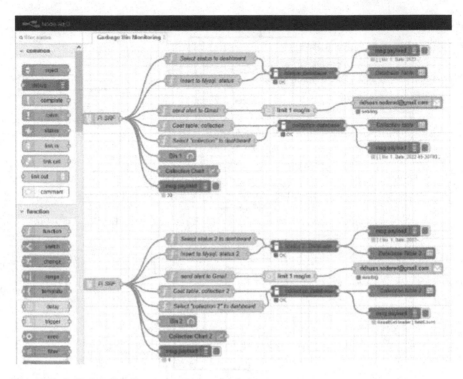

Fig. 4 Node-Red coding flow

Fig. 5 The prototype for a smart dustbin

Fig. 6 Ultrasonic sensor
placement

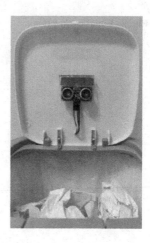

To identify the distance reach the maximum that read by the sensor, msg. payload node is used to show the results as shown in Fig. 7. This will determine whether the dustbin is full or empty. Figure 8 shows the email notification that the system sends to the waste management authority. The email has been given a title to identify which bin is full or empty. When the bin is empty, the system sends out a thank-you message for cleaning the garbage bin. When the bin is almost full, an alert message is sent to notify the authority to clean it up as soon as possible before it is overflown. The Node-Red function allows the system to send the message to the specified email address.

Figure 9 shows the database used to store the data obtained from the sensors. The data of the trash accumulated inside the garbage bin is saved with complete detail such as the date, time, and status. From the waste management authority side, they can fully monitor the data by opening a website provided by the system which is the Node-Red dashboard. The dashboard in Fig. 10 shows the garbage bin status in gauge form, the table data from the database, and chart data. Furthermore, the dashboard may also be viewed on an Android device by typing Raspberry Pi's IP

Fig. 7 Coding flow for
distance is shown in
Node-Red

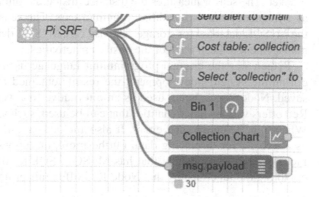

Fig. 8 Email notification

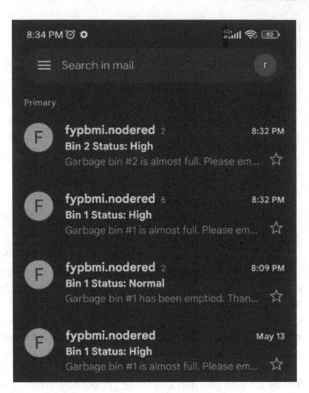

address or hostname and the dashboard will be shown. To view the dashboard, the network must be the same as the one used by Raspberry Pi. The dashboard cannot be displayed if it uses a different network infrastructure.

The project's aims and objectives are successfully achieved. The prototype, which consists of a dustbin coupled to a box at the back, is completed. All of the components used, including the Raspberry Pi, power bank, and ultrasonic sensor, are kept in the box. The ultrasonic sensor is successfully placed at the top of the lid by using a bracket. The sensor measures the distance inside the bin by emitting a sound wave at a frequency that is inaudible to human ears. The sensor then sends the data to the MySQL database for storage. The use of a MySQL database is critical for waste management to keep track of the dustbin's activities.

Python was the original programming language used in this project. However, during the system's development, the results obtained from the sensor cannot be saved. Node-Red is being used as an alternative to developing the project. Node-Red offers various functionalities that can be used, such as sending notifications via WhatsApp, Telegram, and email. It also includes a built-in dashboard that serves as a website for displaying data. Furthermore, the browser-based flow editor also includes a database palette that has MySQL, SQLite, Influxdb, and others. After exploring the functionality that Node-Red offer, this project decided to use email as

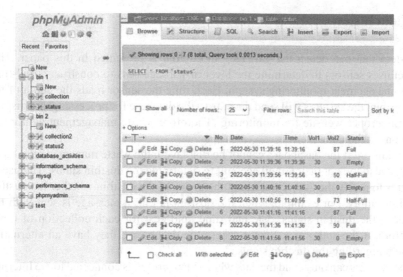

Fig. 9 phpMyAdmin database

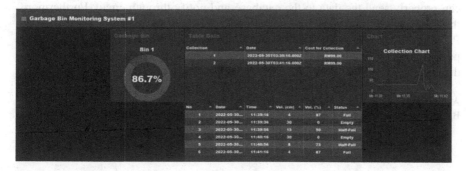

Fig. 10 Node-Red dashboard

a notification to send to waste management. MySQL database is used to store the data and dashboard for display.

In comparison to the previous study, the results of this project are similar to those of [8]. The researcher acquired the results and uploaded them to the ThingSpeak website to show to the waste management authorities [8]. Although the researchers' method differs from that of this project, the result is the same: to verify the dustbin's capacity and notify the waste management authorities.

4 Conclusion

A system that can detect the garbage bin content is presented in this paper. This project uses sensors to determine the level of garbage bins to construct a real-time waste management monitoring system. Each time the sensor reads the depth of the garbage bin, the systems will send the data to the database for processing and display it on the proper website for monitoring. Therefore, waste management authorities can monitor the data.

In addition, if the dashboard cannot be displayed, the waste management authorities can also look at their Gmail account to determine the dustbin status. The use of Gmail is to notify the authorities about the status of the dustbin whether the dustbin is full or empty. For example, when the dustbin is full, a message is sent with the details of the dustbin status, date and time, and cost for each collection of trash. Therefore, if the authorities cannot access the dashboard, they have an alternative for them regarding the dustbin status.

The systems can only send the data when Raspberry Pi is connected to the Internet. Without it, the system cannot function properly. For future recommendations, the system can be equipped with a mini router or mobile broadband. With this, the system can properly send the data to the database and notify through email. Furthermore, the Raspberry Pi is powered by a power bank. When the power is out, the Raspberry Pi is turned off. There are various ways to power the Raspberry Pi such as solar panels, power over ethernet (PoE), batteries, and more. Each has its advantages and disadvantages.

Acknowledgements The authors would like to acknowledge the Communication Section of UniKL BMI for their assistance and support for the success of the project.

References

1. What is the Internet of Things (IoT)? (2022), Oracle.com, https://www.oracle.com/internet-of-things/what-is-iot/. Accessed 31 May 2022
2. R. Wang, Data storage and the Internet of Things (2021), https://phisonblog.com/data-storage-and-the-internet-of-things/. Accessed 8 Nov 2021
3. P. Tambare, P. Venkatachalam, IoT based waste management for smart city. IJIRCCE **4**, 8 (2016)
4. K.V. Kumar, Smart garbage collection bin overflows indicator using IOT. Int. J. Eng. Res. **03**(5), 2288–2290 (2016)
5. K.C. Meghana, K.R. Nataraj, IOT based intelligent bin for smart cities. Int. J. Recent Innov. Trends Comput. Commun. **4**(5), 225–229 (2016)
6. R. Ramly, A.A.B. Sajak, M. Rashid, IoT recycle management system to support green city initiatives. Indones. J. Electr. Eng. Comput. Sci. **15**(2), 1037–1045 (2019). https://doi.org/10.11591/ijeecs.v15.i2.pp1037-1045
7. T.J. Sheng et al., An Internet of Things based smart waste management system using LoRa and Tensorflow deep learning model. IEEE Access **8**, 148793–148811 (2020). https://doi.org/10.1109/ACCESS.2020.3016255

8. M.R. Mustafa, A.K.N.F. Ku, Smart bin: Internet-of-Things garbage monitoring system. ICEESI **140** (2017). https://doi.org/10.1051/matecconf/201714001030
9. R. Tejaswini, R. Ravuri, M.R. Kumar, IOT based smart garbage monitoring system using Raspberry Pi with GPS link (2019), http://ijesc.org/. Accessed 31 May 2022
10. N.A.L. Ali, R. Ramly, A.A.B. Sajak, R. Alrawashdeh, IoT e-waste monitoring system to support smart city initiatives. Int. J. Integr. Eng. **13**(2), 1–9 (2021). https://doi.org/10.30880/ijie.2021.13.02.001

Port Kelang Development Moving Toward Adopting Industrial Revolution 4.0 in the Seaport System: A Review

Kaannappar Chandrasekaran, Anis Farhani Abdul Ghafar, Ahmad Azmeer Roslee, Siti Noor Kamariah Yaacob, Shafu lizam Omar, and Wardiah Mohd Dahalan

Abstract Port Kelang plays an important role in Malaysia's economy especially in trading and transportation of goods due to the presence of three major busiest ports, passenger terminal, and jetties and surrounding industrial areas. Today, technology transformation in industries becomes a hot arising issue by the industry players especially in the maritime sector, particularly in the case considered on the Industrial Revolution 4.0 adoption in the Port Kelang seaport system. The aim of this study was to determine the benefits of the adaption of IR 4.0 in the Port Kelang seaport system meanwhile to identify the challenges faced by terminal and industry players in Port Kelang and to evaluate the implementation of digitalization, automation, and seaport quality at Port Kelang in the current situation and the future. By using focused group

K. Chandrasekaran
Student Development Section, Universiti Kuala Lumpur, Malaysian Institute of Marine Engineering Technology, Lumut, Perak, Malaysia
e-mail: kaannappar.chandrasekaran@s.unikl.edu.my

A. F. Abdul Ghafar
Faculty of Information and Communication Technology, Universiti Teknikal Malaysia Melaka, 76100 Durian Tunggal, Melaka, Malaysia
e-mail: P032110010@student.utem.edu.my

A. A. Roslee
Marine Engineering Section, Universiti Kuala Lumpur, Malaysian Institute of Marine Engineering Technology, 32200 Lumut, Perak, Malaysia
e-mail: ahmadazmeer@unikl.edu.my

S. N. K. Yaacob · S. Omar
Technoputra Section, Universiti Kuala Lumpur, Malaysian Institute of Marine Engineering Technology, 32200 Lumut, Perak, Malaysia
e-mail: sitikamariah@unikl.edu.my

S. Omar
e-mail: shafulizam@unikl.edu.my

W. Mohd Dahalan (✉)
Marine Electrical and Engineering Section, Universiti Kuala Lumpur, Malaysian Institute of Marine Engineering Technology, Lumut, Perak, Malaysia
e-mail: wardiah@unikl.edu.my

© The Author(s), under exclusive license to Springer Nature Switzerland AG 2023
A. Ismail et al. (eds.), *Advances in Technology Transfer Through IoT and IT Solutions*,
SpringerBriefs in Applied Sciences and Technology,
https://doi.org/10.1007/978-3-031-25178-8_8

discussion (FGD) and survey questionnaires, data from the targeted respondent has been collected and analyzed. The questionnaire collects customers' demographic information, industry background, knowledge on IR 4.0, and expectations from the implementation of IR 4.0 in the Port Kelang seaport system. This study intends to provide valuable insight into the implementation of the industrial revolution 4.0 in the seaport system at Port Kelang. A total of 105 questionnaires are collected and the data are analyzed using the statistical package for the social science (SPSS) to determine the benefits of adaption of the IR 4.0 in the Port Kelang seaport system and to identify the challenges faced by terminal and industry players in Port Kelang. The result found a strong relationship between the terminal facilities and industry players' satisfaction. On the other hand, it can improve the understanding of the current operation system in the seaport at Port Kelang and is able to evaluate the implementation procedure of IR 4.0.

Keywords Port Kelang · Industrial revolution · IR 4.0 · Seaport system

1 Introduction

Port Kelang is a well-known city and Malaysia's largest port, located on the west coast of Selangor along the river Kelang in the Straits of Malacca. Initially, Port Kelang was named and called Port Swettenham. The Port of Swettenham, which was created very recently compared to Penang and Singapore and has only been in operation since 1901, has become one of the top three ports in Malaysia, and thus it plays a crucial role in the development of central Malaysia from Ipoh to Malacca and western Pahang in the east [1]. Port Kelang plays an important role in Malaysia's economy especially in trading and transportation of goods due to the presence of three major busiest ports, passenger terminal, and jetties and surrounding an industrial areas. Furthermore, Port Kelang is also situated at a strategical point, thus Port Kelang is nearby Kuala Lumpur, the federal capital, is located 37 km east-northeast as well as Kuala Lumpur International Airport (KLIA), and is accessible by road and rail.

Three major Port terminals were playing an essential role in Port Kelang in maritime operation and transportation industry in Westport, Northport, and Southpoint. Besides that, Port Kelang Free Zone (PKFZ) and private jetties such as Lunar Jetty, Pulau Ketam Jetty, and Asa Niaga also contribute a lot to Port Kelang's development and the nation's economy as well. Hereby, Port Kelang Authority (PKA) acts as a key factor in administrating and monitoring management and operation procedure of port terminals, PKFZ, and private jetties in Port Kelang. Port Kelang has been handling almost 50% of cargo volume globally and more than 70,000 ships passing through Port Kelang per annum as well as providing over 20% of ships passing through the Straits of Malacca [2]. In the year 2002, Port Kelang was placed top 15th port globally and is able to maintain the ranking position until now. Technology transformation in port operation is the main driver for the current transition in

Port Kelang. Technology transformation in Port Kelang is relatable to the industrial revolution. Industrial Revolution 4.0 in the maritime industry is stated as Maritime 4.0. Maritime 4.0 defines the focus on information and communication technology (ICT) and digitalization and information-oriented industry [3].

The main issue faced in the Port Kelang seaport is human error, and port facility services are common issues brought up during the traditional operation. According to that, accidents happening within port premises, especially vessel, cargo, and human, includes that. Cargo and vessel movements within the port limits are also arising issues faced by industry players especially shipping agents, freight or forwarding agents, and transportation companies. Labor issues are also considered as serious points that arise from terminals. Labor protests for salary and lack of operational knowledge are highlighted based on labor issues in terminals. Increasing labor costs is prolonging the payback period of investment and reduce the skilled workforce and operational efficiency [4]. Next, the port administration system found deficiencies since the present time of many vendor system providers. According to the situation, data varies and creates a chaotic situation among port system users. The aim of this research is to determine the level of customer satisfaction toward the Kuala Perlis ferry terminal. The objective of this study is to evaluate the current status of customer satisfaction with ferry operations at Kuala Perlis and to investigate the influence of terminal facilities toward customer satisfaction at the Kuala Perlis Terminal.

The main results of this study will also benefit the Port Kelang seaport management as they can use the information on industrial players' satisfaction to improve the seaport facilities according to the introduction of new technology based on IR 4.0. In addition, Malaysia's economy should be increased to build the country's development. Next, the management and operation process of the seaport in Port Kelang goes on faster and creates a comfortable environment for customers to conduct their business in Port Kelang. According to that, customers will be satisfied with the management and services that have been provided. Finally, it will provide more information about seaport operation in Port Kelang and initiatives of the seaport in Port Kelang adopting IR 4.0 in its operational and management system as well this research can be a reference to future researchers.

This study investigates the preparedness of Port Kelang management in the adaptation of this new IR 4.0 concept, benefits, and limitations as well as the prediction of Port Kelang in port operation in future. The purpose of this research is to investigate the advantages of the adaptation of IR 4.0 in the Port Kelang seaport system. Besides that, it identifies the challenges faced by terminal and industry players in Port Kelang and assesses the implication of digitalization and seaport quality at Port Kelang in the current situation and the future.

2 Analysis of the Previous Works

2.1 Industrial Revolution

The first revolution began in 1784 in European countries. The first idea was to use water and steam power engines spread all over the world. It paved the way for the world of the revival of agriculture and feudalism to modern manufacturing [5]. The second revolution was the creation of large-scale use of electric power. It was then developed into the third revolution using electronics and information technology to automate production [6]. Finally, the current fourth industrial revolution (IR 4.0) is formed which focuses more on digital technology information. The three important elements in the era of IR 4.0 are the differences between the physical, digital, and biological realms.

The primary motive for implementing Industrial Revolution 4.0 is to enhance the availability of data with accurate results by introducing new technology systems accordingly such as Internet of Things (IoT), smart sensors, advanced robotics, location detection, big data analytics, cloud computing, 3D printing, and augmented reality [7]. However, it requires the latest technology telecommunication network system known as the 5th generation system (5G) to operate smoothly without any lag. Industrial Revolution 4.0 in the maritime industry is stated as Maritime 4.0. In accordance, this idea was initially coined by Germany in the European parliament. Currently, the idea is implemented in many ports in the world, and there are certain developing countries that are moving toward it; Malaysia is one of them in that list.

Maritime 4.0 defines the focus on information and communication technology (ICT) and digitalization and information-oriented industry [3]. Digitalization and automation concept plays a vital role in the implementation of Maritime 4.0 in ports. Digitalization is explained as the transfer of data or operational processes that we previously created manually to a computer or to a digital environment. Meanwhile, the primary motive makes sure the process and operation go on faster and data access and resource management will simplify especially in data collection, storage, and use of large and diverse information [8]. Under the point of digitalization where it comes to the blockchain system and digital twin system to enhance the seaport management and operation system to be faster, obtaining accurate data and real-time data by using the latest technology is important [9]. This introduction of new technology basically improves the port operation as well as the production rate of the seaport and the economy of the nation accordingly. According to the situation, there was the current transition technology used by terminals in Port Kelang such as Westport and Northport that is still using the 4G technology which was still counted as IR 3.0. They were monitoring the cargo movement and port operation by using CCTV installation. Besides that, terminals in Port Kelang were awarded as green ports for their electric rubber tire gantry (RTG) and paperless system for the gate and cargo clearance in port operation. This study will identify the preparedness of terminals in Port Kelang in adopting IR 4.0 elements in their operation as well in their management.

2.2 Seaport System

A port can be defined as a harbor region or geographical area that can accommodate a large number of boats and vessels for berthing and anchorage facilities [10]. Besides that, a port is also defined to carry out transferring of people or cargo from ship to shore and vice versa while also allowing for continuous or periodic shipment operations. Unlike early ports, which were mostly just harbors, however, modern ports now act as multimodal distribution hubs with transportation links via sea, river, canal, road, train, and air [11]. Ports are also classified as comprising terminals such as container terminals and conventional terminals (break bulk, dry bulk, and liquid bulk). There are several types of ports such as seaport, passenger port, fishing port, inland port, warm water port, and dry port. All above-mentioned ports play a crucial role and provide a good revenue return as well as contribute to our nation's economy accordingly.

Port Kelang consists of very important seaports in Malaysia such as Westport, Northport, and Southpoint. Besides that, Port Kelang also plays a vital role as a passenger port for example Pulau Ketam jetty, Asa Niaga jetty, Boustead Crusie, and Lunar jetty. To this place, we are able to understand the current traditional port operation as well as are able to determine the importance of IR 4.0 adoption in benefits the port operation and management.

3 Methodology

The study was surveyed and analyzed at the seaport in Port Kelang which is positioned on the west coast of Peninsular Malaysia. The survey is conducted throughout Westport, Northport, and Southpoint principles as well as industry players related to the maritime industry such as shipping agents, forwarding agents, freight forwarders, and transportation entities, port authority, Marine Department, Customs Department, and Immigration Department located in Port Kelang as well as Ministry of Transport (MOT). Targeted respondents were set as 25 in order to obtain a precious end result. The purpose of this research is mainly to learn and gain knowledge information about the adaption of IR 4.0 implementation in the current traditional operational system and to determine the most critical sustainable factor (CSF's) needed to make IR 4.0 a reality in the port system of Port Kelang [3].

Furthermore, it is noticed that in comparison to the global trend, there are few studies examining the present status of seaports and CSFs to enable IR 4.0 in the seaport system, and qualitative studies through a focused group discussion (FGD) [12]; similarly, providing Questionnaires is needed to comprehend the current reality in the Malaysian context especially seaports in Port Kelang since it is seen as an important key factor then delate and its role to the nation. An FGD is a frequent technique for novel fact research and discovery, as well as exploratory grounded investigations [13]. Besides that, FGD is a good way to get a comprehensive grasp of a new phenomenon [14]. Our focus group facilitated an open and in-depth discussion of the researcher's chosen topics and looked at the current state of CSFs for IR 4.0 implementation.

4 Analysis of the Results

The FGD lasted three (3) hours and featured a group of ten (10) specialists. The experts were carefully picked, using the FGD technique, based on their responsibilities as decision-makers and their recognition as experts in the field. We have conducted research on FGD by questioning the implementation status in accordance with the Malaysian situation, as well as the CSFs required to carry out sustainable IR 4.0 in the port industry; to identify the implementation status of IR 4.0 more clearly, video and audio recordings were used to record FGD data. The audio and video recordings were transcribed in their entirety and the data was analyzed for codes and themes. This study uses the 'thematic analysis' as its main tool. Thematic analysis is a qualitative research technique that has been applied to a variety of topics and research. During the thematic analysis, six phases were implemented such as (1) familiarizing oneself with one's data, (2) creating initial codes, (3) finding themes, (4) reviewing themes, (5) defining topic nomenclature, and (6) compiling a report.

Upon analysis of the study's findings, the following conclusions were reached where six CSFs were determined. Those are environmental protection, social equity and culture, economic value, operational value and technical quality, communication and cooperation, and political and legal parameters. As per analysis, IR 4.0 basically provides huge benefits to the industry such as improving the productivity rate, reducing lead time, increasing competitive skills, and improving the profits of the terminal. However, to achieve it there are several challenges faced by terminals as well as industry players such as adaptability to technology, hiring a skilled workforce, social liability, and high investment. At these moments, CFS identified plays an important role in pertaining to the process of adaptation of IR 4.0 in Port Kelang.

5 Conclusion

Based on the results from this study, it is found that implementation of IR 4.0 in Port Kelang brings a lot of benefits to the maritime industry as well as improves the economic inflow of our nation. Several important factors need to be considered in implementing IR 4.0. Social and cultural values as well as environmental, economic, and technical quality, as well as political and legal cooperation, should be considered by seaport users in Kelang Port. Furthermore, more research can be done to investigate and verify the relationship between the elements that influence the use of IR 4.0 in seaports, especially in new projects in Port Kelang such as the third port in Pulau Carrey island and the Westport new phase. In general, the path to the objective has been examined, and investment and clear policy-making methods must be established ahead of time in order to arrive at the destination [15].

References

1. N.H. Mohd Salleh, M. Selvaduray et al., Adaptation of Industrial Revolution 4.0 in a seaport system. Sustainability (13), 10667 (2021)
2. L. Sommer, Industrial revolution-industry 4.0: are German manufacturing SMEs the first victims of this revolution? J. Ind. Eng. Manag. **8**(5), 1512–1532 (2015)
3. M.A. Islam, A.H. Jantan et al., Fourth industrial revolution in developing countries: a case on Bangladesh. J. Manag. Inf. Decis. Sci. **21**(1), 1–9 (2018)
4. A. Bottalico, Towards a common trajectory of port labour systems in Europe? The case of the port of Antwerp. Transp. Policy **8**(2), 311–321 (2020). https://doi.org/10.1016/j.cstp.2019.12.003
5. E. Simmons, G. McLean, Understanding the paradigm shift in maritime education: the role of 4th Industrial Revolution technologies: an industry perspective. Worldw. Hosp. Tour. Themes **12**(1), 90–97 (2020)
6. M. Xu, J. David, S.H. Kim, The fourth industrial revolution: opportunities and challenges. J. Financ. Res. **9**(2), 90–95 (2018)
7. B.P. Sullivan, S. Desai, J. Sole et al., Maritime 4.0. Opportunities in digitalization and advanced manufacturing for vessel development. Procedia Manuf. **42**, 246–253 (2020). https://doi.org/10.1016/j.promfg.2020.02.078
8. A.A. Shahroom, N. Hussin, Industrial revolution 4.0 and education. Int. J. Acad. Res. Bus. Soc. Sci. **8**(9), 314–319 (2018)
9. O.P. Brunila, V.K. Hyrkki et al., Hindrances in port digitalization? Identifying problems in adoption and implementation. Eur. Transp. Res. Rev. (2021). https://doi.org/10.1186/s12544-021-00523-0
10. K. Button, A. Chin, T. Kramberger, Incorporating subjective elements into liners' seaport choice assessments. Transp. Policy **44**, 125–133 (2015)
11. S. Lim, S. Pettit, W. Abouarghoub et al., Port sustainability and performance: a systematic literature review. Transp. Res. Part D: Transp. Environ. **72**, 47–64 (2019). https://doi.org/10.1016/j.trd.2019.04.009
12. F. Moretti et al., A standardized approach to qualitative content analysis of focus group discussions from different countries. Patient Educ. Couns. **82**(3), 420–428 (2011). https://doi.org/10.1016/j.pec.2011.01.005
13. L.D.R. Kurtz et al., Skill building: assessing the evidence. Psychiatr. Serv. **65**(6), 727–738 (2014)
14. D. Stokes, R. Bergin, Methodology or 'methodolatry'? An evaluation of focus groups and depth interviews. Qual. Mark. Res. **9**(1), 26–37 (2006). https://doi.org/10.1108/13522750610640530
15. R. Ryacudu, I.N. Putra, S.A. Purwantoro, Strengthening total people's defense and security system in the Industrial Revolution Era 4.0 to face the threat of sixth generation war. J. Inov. Pertahanan Keamanan **7**(2), 191–204 (2021)

Investigating the Performance of Optimizing the Convolutional Neural Network in Detecting Malware Attack

Roziyani Rawi, Muhammad Hazim Najmi Hamka, Husna Sarirah Husin, and Noormadinah Allias

Abstract Malware is one of the major issues in cybersecurity and among computer users. It has caused severe loss to businesses, organizations, and people. Therefore, this research aims to detect malware using the convolutional neural network (CNN) algorithm. Chi-Square has been used and has selected twenty important features as input for the machine learning models. In addition, random and grid searches have been used to optimize the performance of the CNN and determine the best hyperparameter on the algorithm. The experiments show that CNN has outperformed Neural Network's performance in terms of accuracy and precision. Meanwhile, the CNN has random search accuracy and precision; thus, we conclude the randomized search algorithm produced a good prediction result with CNN for large datasets.

Keywords Malware detection · Machine learning · Hyperparameter tuning · Convolutional neural network

R. Rawi (✉) · M. H. N. Hamka · H. S. Husin
Universiti Kuala Lumpur, Malaysian Institute of Information Technology, 50250 Kuala Lumpur, Wilayah Persekutuan, Malaysia
e-mail: roziyani@unikl.edu.my

M. H. N. Hamka
e-mail: hazim.hamka@s.unikl.edu.my

H. S. Husin
e-mail: sarirah@unikl.edu.my

H. S. Husin
Centre for Women Advancement and Leadership, Universiti Kuala Lumpur, 50250 Kuala Lumpur, Wilayah Persekutuan, Malaysia

N. Allias
School of Computing and Informatics, Albukhary International University Alor Setar, Kedah, Malaysia
e-mail: noormadinah.allias@aiu.edu.my

© The Author(s), under exclusive license to Springer Nature Switzerland AG 2023
A. Ismail et al. (eds.), *Advances in Technology Transfer Through IoT and IT Solutions*,
SpringerBriefs in Applied Sciences and Technology,
https://doi.org/10.1007/978-3-031-25178-8_9

1   Introduction

Malware can be defined as software that is designed to disrupt, harm, or gain unauthorized access to a computer system or network. It can be divided into adware, backdoor, bot, downloader, launcher, ransomware, rootkit, spyware, Trojan, virus, and worm [1]. Each form of malware has its method of wreaking havoc, and most of them depend on user action in some way. Effectively shielding machines from malware infections has become critical for businesses, organizations, and people. Many methods have been introduced for detecting malware, such as signature-based detection, heuristic-based detection, and specification-based detection. However, these methods have some limitations. Current commercial antivirus (AV) software cannot provide the level of security required. Thus, machine learning has been seen as a potential method that can be used to detect malware.

Initially, the convolutional neural network (CNN) has been widely used for images. Therefore, the main aim of this research is to investigate the performance of CNN in detecting malware and compare it with traditional machine learning algorithms, Random Forest and Neural Networks. However, all the machine learning algorithms mentioned above share a similarity whereby they have hyperparameters that need to be tuned to maximize their performances. As a result, this research intends to investigate the performance of hyper-tuning CNN on malware detection. The Grid search and Randomized search have been selected. The result will be compared with other machine learning algorithms, including random forest and neural networks, in terms of precision, recall, and accuracy.

Meanwhile, according to a survey conducted by Bazrafshan et al. [2], three methods for detecting malware were considered: signature-based, behaviour-based, and heuristic-based. Signature-based methods are more efficient and faster than other methods since they leverage patterns from diverse malware to identify them. Because these fingerprints are generally recovered with high sensitivity because they are unique, detection methods that utilize them have a low error rate [3]. Malware detection techniques based on behaviour examine a program's behaviour to determine whether it is malicious or not. Heuristic malware detection systems employ data mining and machine learning approaches to learn an executable file's behaviour. An example of this is a study by [4] that gathered traces of app behaviour on Android and separated them into two groups to distinguish benign apps from malware-infected apps clearly.

According to [5], most existing malware classification approaches focus on improving precision rather than detection speed, and real-time malware detection is still in progress. Thus, the author proposed a quick portable executable (PE) malware detection method using Chi-Square as an effective feature filtering approach. The method will choose the most significant APIs and TPFs. In addition, the Phi coefficient divides the features into separate subsets depending on their importance. From the testing, 98% accuracy is obtained, and files (malware or benign) can be categorized within 0.09 s.

Research performed by [6] discusses using deep learning architectures such as CNN, DNN, and RNN to be compared with a classical machine learning algorithm. The authors stated that malware detection using two convolutional layers in CNN with LSTM produces the highest results among the deep learning architectures and classical machine learning. The authors' use of the best hyperparameter tune stated that the CNN is best to be used with softmax activation, Adam optimizer, and binary cross-entropy as a loss function.

The authors [7] proposed malware detection using 1D convolutional neural networks. Thus, using a 1D convolutional layer is the key to creating a CNN model in malware detection, and a result obtained from the experiments are very high. A study by [8] shows that random forest outperforms deep neural networks with opcode frequency as a feature.

The author of [9] proposed a CNN-based architecture to classify malware samples. They convert malware binaries to grayscale images and train a CNN for classification. Experiments on two challenging malware classification datasets, Malimg and Microsoft malware, have shown the proposed methods achieved better performance, where 98.52% and 99.97% accuracy have been achieved on the Malimg and Microsoft datasets, respectively.

2   Methodology

2.1   Proposed CNN Model Framework

Figure 1 shows the CNN malware detection framework used in this experiment. While, Fig. 2 shows the proposed CNN model framework which is divided into the feature extraction/learning phase and the classification phase. The input layer represented by the neurons will provide the initial data into the system for artificial neurons process. The input is the dataset of PE malware and benign files that will be extracted into the convolution layer that contains 32 filters with a size of 3×3 each. Here, the activation of ReLU was used as it has become the default activation function for many types of neural networks because it is easier to train and often achieves better performance [7]. Then, a collection of learnable filters is used in the convolutional layer. Specified features or patterns in the input are detected using a filter. The convolutions 1D are used to look for features in the sequences [7].

Another convolution layer was added to improve the feature learning process and the detection performance. Next, the features will be extracted to the max pooling layer, a feature map pooling procedure that estimates the maximum or most significant value in each patch. Then, another max pooling layer was added to the proposed CNN model framework. Before a fully connected layer was built, LSTM, which stands for Low Short-Term Memory, was used to process and generate predictions given data sequences; it is meant to function differently than a CNN. Based on

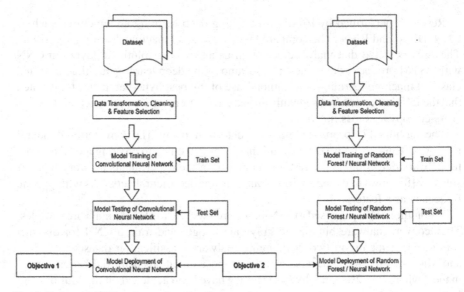

Fig. 1 Proposed CNN malware detection framework

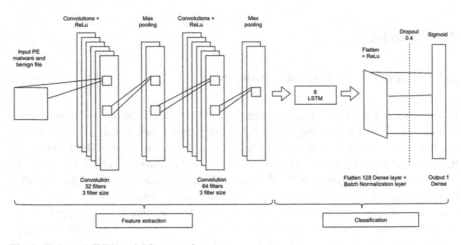

Fig. 2 Proposed CNN model framework

research by [6], the uses of the LSTM are proven to increase malware detection performance.

Adam optimizer with a learning rate of 0.01 and binary cross-entropy is used based on the research work in [6]. Then, three flattening layers are applied to convert the pooled feature map to a single column passed to the fully connected layer, and dense gives the CNN a fully connected layer. Dropouts used in flattening layer functions will prevent the CNN from overfitting. Sigmoid activation acts as a classifier in the classification phase on the output layer. Lastly, the output layer will produce the detection results after completing all the learning and classification phases.

3 Results and Discussion

This research phase evaluates and discusses the best hyperparameter algorithms for detecting malware using the CNN algorithm.

3.1 Experiment to Detect Malware Using Hyperparameter Tuning CNN

The PE malware and benign file dataset were used to perform this experiment using the feature selection technique Chi-Square to increase the efficiency of the machine learning:

(a) The system was trained with PE malware and benign file dataset, whereby 70% of the dataset belonged to the training dataset, and 30% of the dataset belonged to the testing dataset from 19,611 of the total malware and benign files.

(b) About 13,728 files of PE malware and benign file were used for training, and 5883 files were used for testing purposes, randomly selected from the dataset. Later, the training dataset was used to build a model by training the CNN algorithm before being tested with the testing dataset.

(c) The Chi-Square technique was used to select important features by choosing features that are strongly dependent on the response when selecting features.

(d) All the experiments were conducted then, and the hyperparameter was tuned using the grid search and randomized search to find the best parameter of the algorithm.

(e) Finally, the best hyperparameter model tuning will be tested using a test dataset to evaluate the effectiveness of machine learning in detecting malware. The analysis of the results is discussed in Sect. 3.2.

3.2 Result Analysis

This section discusses malware detection results using hyperparameter tuning convolutional neural networks.

Several observations are made based on the experimental result in Table 1. This observation is discussed by analyzing the result of detection performance and the best hyperparameter tuned method. While the results in Table 4 show the test dataset's precision, recall, and accuracy after the testing process, Fig. 3 shows the performance of the detections over parameter settings.

Table 1 Result of malware
detection using
hyperparameter tuning
convolutional neural network

Hyperparameter tune method	Accuracy (%)	Precision (%)	Recall (%)
Grid search	88.34	99.46	84.62
Randomized search	97.62	98.84	97.92

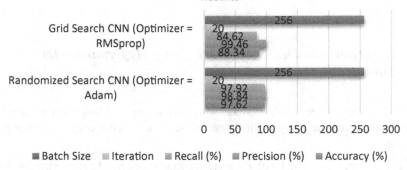

Fig. 3 Parameter setting versus performance detection

The result obtained from the experiment in Table 1 and Fig. 3 shows that the randomized search championed CNN's best detection performance when tested on massive datasets compared to the grid search [10]. Besides, the optimizer used in the algorithm was the main factor in boosting the detection performance. Meanwhile, based on Table 1 and Fig. 3, it is observed that the performance of randomized search on CNN achieved the highest detection result with the iteration (epochs) = 20, batch size = 256, and optimizer = Adam. However, in a grid search, the parameter was set the same as a randomized search, but the optimizer used was RMSprop. From the observation, by using the Adam optimizer, the detection accuracy increased 13.3%, 9.28% increment in the recall, but the grid search has higher performance in terms of precision. Thus, it could be related that the optimizer's use in this algorithm affects the malware detection performance in this experiment.

3.3 Comparison Performance of Hyperparameter Tuning CNN with Other Machine Learning

The procedure to conduct this experiment is the same as the experiment using the CNN algorithm, including the feature selection technique with the same dataset. The section discusses the performance comparison between the CNN with random forest

Table 2 Result of machine learning performance without feature selection and hyperparameter tuning

Machine learning	Accuracy (%)	Precision (%)	Recall (%)
Random forest	98.8	99	98
Neural network	74.86	74.86	100
Convolutional neural network	63.53	99.56	51.18

and neural network. Each selected machine learning will be tuned to find its best parameter setting via grid and randomized search.

From Table 2, the result of each machine learning performance was evaluated to determine the standard performance before hyperparameter tuning and without using the feature selection technique. According to [11], the random forest algorithm had the best advantage since the dataset size is large. The performance of the random forest was the highest among the others because of its outstanding features like variable importance measure, OOB error detection, proximity among the feature, and handling of imbalanced datasets. The performance of CNN was the lowest among these machine learning except for the precision score, which was the highest indicating that 99.56% of results retrieved by a search were relevant.

As dictated, Table 3 shows the improved performance of the CNN algorithm in detecting malware. The feature selection technique using Chi-Square helps the algorithm to select only important features, where k = 20, determining that the algorithm will process only the best 20 features in the dataset. So, it can be related that the feature selection process using Chi-Square was essential in improving the efficiency of the algorithms.

Meanwhile, Table 4 shows that each machine learning performance to detect malware increased. The overall machine learning results were increased because the grid search suggested the best parameter for this experiment. However, the process of searching all possible combinations prior to suggest the best parameter will cause more time consuming.

In addition, observations made from Table 5 indicate the best results of this machine learning using randomized search hyperparameter tuning. Due to the large dataset size, it is possible that the hyperparameter tuning using randomized search was the highest. The CNN algorithm performed better than the neural network when

Table 3 Result of machine learning performance with feature selection

Machine learning	Accuracy (%)	Precision (%)	Recall (%)
Random forest	98.9	99	98
Neural network	74.92	74.92	100
Convolutional neural network	88.28	86.47	99.92

Table 4 Hyperparameter tuning using grid search machine learning performance with feature selection

Machine learning	Accuracy (%)	Precision (%)	Recall (%)
Random forest	98.8	99	97
Neural network	96.26	97.01	98.03
Convolutional neural network	88.34	99.46	84.62

Table 5 Hyperparameter tuning using randomized search machine learning performance with feature selection

Machine learning	Accuracy (%)	Precision (%)	Recall (%)
Random forest	98.9	99	98
Neural network	96.60	97.23	98.26
Convolutional neural network	97.62	98.84	97.92

the best hyperparameter values were used to detect malware. Thus, the best parameter setting can provide the maximum performance for the CNN algorithm to detect malware, besides its efficiency for image analysis without hyperparameter tuning.

4 Conclusion

This research was conducted to classify malware and benign file detection in terms of accuracy, precision, and recall. This research focused on implementing a malware detection method using hyperparameter tuning in the CNN algorithm. The findings from this research showed that the CNN algorithm can still detect malware without the feature selecting technique but not with excellent results. This factor was influenced by using a feature selection technique, which helped to choose the most important feature for the algorithm to make a detection. The comparison results of the CNN algorithm with Random Forest and Neural Network show that this research framework produced better detection performance than the neural networks. The performance has been evaluated based on accuracy, precision, and recall.

In conclusion, this research has answered the problem statement and achieved the objectives. However, the time taken for hyperparameter tuning to find the best parameter is high due to the large dataset being processed. For future works, instead of investigating the performance of CNN on malware, we would like to examine the performance of CNN in detecting Distributed Denial of Services (DDoS), as this attack is becoming one of the most common attacks in the cyber world.

Acknowledgements We would like to express our sincere gratitude to all members involved in this study directly and indirectly. Many thanks to Universiti Kuala Lumpur, Malaysian Institute of Information Technology (UniKL MIIT) for the funding and support.

References

1. D. Gibert, C. Mateu, J. Planes, The rise of machine learning for detection and classification of malware: research developments, trends and challenges. JNCA **153**, 102526 (2020)
2. Z. Bazrafshan, H. Hashemi, S.M.H. Fard, A. Hamzeh, A survey on heuristic malware detection techniques, in *The 5th Conference on IKT* (IEEE, 2013), pp. 113–120
3. P. Gutmann, The commercial malware industry, in *DEFCON Conference* (2007)
4. I. Burguera, U. Zurutuza, S. Nadjm-Tehrani, Crowdroid: behavior-based malware detection system for Android, in *Proceedings of the 1st ACM Workshop on SPSM'11* (2011), pp. 15–26
5. M. Belaoued, S. Mazouzi, A chi-square-based decision for real-time malware detection using PE-file features. JIPS **12**(4), 644–660 (2016)
6. R. Vinayakumar, M. Alazab, K.P. Soman, P. Poornachandran, S. Venkatraman, Robust intelligent malware detection using deep learning. IEEE Access **7**, 46717–46738 (2019)
7. A. Sharma, P. Malacaria, M.H.R. Khouzani, Malware detection using 1-dimensional convolutional neural networks, in *2019 IEEE EuroS&PW* (IEEE, 2019), pp. 247–256
8. H. Rathore, S. Agarwal, S.K. Sahay, M. Sewak, Malware detection using machine learning and deep learning, in *International Conference on BDACS* (Springer, Cham, 2018), pp. 402–411
9. M. Kalash, M. Rochan, N. Mohammed, N.D. Bruce, Y. Wang, F. Iqbal, Malware classification with deep convolutional neural networks, in *2018 9th IFIP International Conference on (NTMS)* (IEEE, 2018), pp. 1–5
10. P. Worcester, A comparison of grid search and randomized search using scikit learn (2019), https://medium.com. Accessed 24 Aug 2022
11. V.M. Herrera, T.M. Khoshgoftaar, F. Villanustre, B. Furht, Random forest implementation and optimization for Big Data analytics on LexisNexis's high performance computing cluster platform. J. Big Data **6**(1), 1–36 (2019)

Proposed Design Principles of Malay Sign Language Mobile Application for Hearing-Impaired Alpha Generation Based on Nielsen and Molich's Design Guidelines: Validation Through Prototyping Method

Ermiera Shafika Mokhtar, Azizah Che Omar, and Nurulnadwan Aziz

Abstract Researchers commonly use the heuristics evaluation method to verify that the system created is usable. The evaluation based on Nielsen and Molich's design guidelines (NMDG) is the widely used methodology. This design guideline however has been reformed and extended in the past by researchers that consider the circumstances of designing mobile applications that may evoke users' cognitive abilities. However, the circumstances of the hearing-impaired (HI) alpha generations are not considered. Therefore, this research focuses on proposing the design principles for Malay Sign Language (MSL) mobile applications that consider the context of HI learners. The proposed design principles were validated using a prototype method covered in-depth in this paper to ensure their dependability.

Keywords Mobile human–computer interaction · Interaction design · Assistive technology · User-centered design (UCD) approach · User experience

E. S. Mokhtar (✉) · A. C. Omar
School of Multimedia Technology and Communication, Universiti Utara Malaysia, 06010 Sintok, Kedah, Malaysia
e-mail: ermiera_shafika_m@ahsgs.uum.edu.my

A. C. Omar
e-mail: co.azizah@uum.edu.my

N. Aziz
Department Research and Industrial Linkages, Universiti Teknologi MARA Cawangan Terengganu, 23000 Dungun, Terengganu, Malaysia
e-mail: nurulnadwan@uitm.edu.my

Faculty of Business and Management, Universiti Teknologi MARA Cawangan Terengganu, 23000 Dungun, Terengganu, Malaysia

1 Introduction

Hearing impairment is a sensory issue that is relatively common in children [1]. Hearing-impaired (HI) alpha generations can be defined as children born after 2010 with difficulties using their hearing organs [2]. The World Health Organization (WHO) classifies hearing impairment in the better ear as mild, moderate, severe, or profound [3]. Additionally, the HI alpha generations may suffer from cognitive problems brought on by hearing loss, including issues with comprehension and memorization of learning materials, including sign languages [4].

Malay sign language (MSL) was developed based on Malay-based words and dialects to facilitate communication between HI individuals in Malaysia and society [5]. Numerous MSL mobile applications have been designed and constructed to help HI individuals interactively learn sign languages, especially the alpha generations. Research by [6], however, reveals that the existing MSL mobile apps do not meet the users' requirements and satisfaction. Additionally, according to [7], the current MSL mobile applications fail to stimulate the cognitive abilities of HI alpha generations. Therefore, it is essential to make sure that the created mobile application is thoroughly evaluated. To ensure the mobile applications meet the users' expectations, [7] mentioned that the mobile applications need to be assessed by the users or the experts to solve the usability issues through usability evaluations.

Usability heuristic evaluation is a method that allows experts to validate that the system interface is used in compliance with a set of principles [8]. The NMDG is the most well-known heuristic in the literature, according to [9], since it is a time- and money-saving strategy. The authors of [7] also encouraged hybridizing the NMDG while developing mobile applications, especially for the HI alpha generations, as it could stimulate their cognitive abilities. However, the existing NMDG that has been expanded by [10] does not consider the context of HI individuals. Thus, sixteen design principles of the MSL mobile applications were constructed based on the expanded NMDG. The proposed design principles were validated through a prototyping approach to ensure their reliability and dependability.

Validation is evaluating and ensuring that the product and system developed can fulfill their specifications and serve their intended functions [11]. In this paper, the proposed design principles were validated through the prototyping method to guarantee that they are reliable for future researchers to refer to when designing the MSL mobile application for HI learners, specifically the alpha generations. This interprets that to ensure that the proposed design principles are useful and reliable, they must be visualized into a working prototype. The proposed design principles are validated once the prototype has been successfully designed and developed (as desired). A semi-working prototype was designed, and the methods involved were discussed in the following sections.

2 Methodology

The prototype was created and designed in Malay as requested by the users' representatives, teachers of the HI alpha generations. The users' representatives were involved in designing the prototype to ensure that the developed prototype could meet the users' needs and satisfactions based on the proposed design principles [12]. This method is also known as the user-centered design (UCD) approach. The rationale for involving the users' representatives instead of the actual users is that the actual users are children with hearing impairment. Thus, they require a more straightforward and more naturalistic method of getting their feedback.

Next, the prototype was designed with the free online software named 'Marvel Apps.' The rationale for choosing this software is that it is easy to use and cost savings since it is free. Besides, it enabled the designers to design the prototype interactively in various device sizes and types by providing ready-made templates. Thus, it can save the designers time in designing the prototype. To ensure that the learning materials adhere to the Malaysian education standard that the Ministry of Education Malaysia has enforced, the teachers of the HI alpha generations prepared the video learning provided in this prototype. The videos were shot, edited, and posted on the school's YouTube site; they can be shared and used for educational purposes without restriction. In this study, the prototype was designed based on the sixteen proposed design principles. Accordingly, the design of the prototype was extensively discussed in the following sections.

3 Results and Discussion

The prototype was designed and developed based on the design principles that have been constructed and proposed based on the NMDG that has been expanded by [10]. The descriptions of the design and the NMDG were discussed as follows.

3.1 Usability Heuristics NMDG 1: Visibility of System Status

Design Principle 1: All screens should be provided with instructions for the HI alpha generations to be aware of their current states.

The prototype of the MSL mobile application provides the instructions and the title on all screens. It was written in plain text form with different colors. This is to differentiate between the title of the learning categories and instructions to prevent the HI alpha generations from being confused. The instructions and the titles were also written in Bahasa Malaysia as requested by the teachers of the HI alpha generations. Figure 1 depicts the sample of the screen.

Fig. 1 Instruction provide

3.2 Usability Heuristics NMDG 2: Match Between System and Real World

Design Principle 2: The sign language learning contents provided should follow the local education standards for the HI alpha generations.

This prototype was designed based on the 'Kod Tangan Bahasa Malaysia (KTBM).' It is a formal and standard sign language used by the teachers of the HI alpha generations in teaching sign language at schools. Besides, the contents of this prototype were adopted from the textbooks and videos provided by the teachers, as shown in Fig. 2. This ensures that this prototype follows the local education standard and can stimulate cognitive ability among the HI alpha generations.

Fig. 2 Learning content

3.3   Usability Heuristics NMDG 3: User Control and Freedom

Design Principle 3: The MSL mobile applications should allow the HI alpha generations to leave the applications anytime.

This prototype provides two exit functions (home button: yellow; exit button: red with 'x'), as shown in Fig. 1. The home button enables the users to leave their current tasks and leads them to the home page of this application for them to choose other activities to be carried out. The rationale for providing only a home button for exit on all screens except for the home page is to motivate the users to explore more on the MSL mobile applications through the various categories offered on the homepage. However, the 'X' button was provided in red color on the homepage. This enables the users to leave the MSL application and stop carrying out any tasks. Thus, this prototype offers flexibility to the HI alpha generations since they can leave the applications anytime.

3.4   Usability Heuristics NMDG 4: Consistency and Standard

Design Principle 4: The quizzes or exercises provided must be coherent with the contents of the video learning.

This prototype offers additional exercises to assess the cognitive ability among the HI alpha generations. The contents of the exercises are similar to the videos learning provided. This is to avoid the user's loss and confusion. The exercises were given in static images, and the users needed to click on the button with the correct answers. Figure 1 shows the sample of the exercises.

Design Principle 5: The fonts used for the texts in the MSL mobile applications should follow the local education standard font types and sizes.

As for the font styles and sizes, this prototype uses the Comic Sans font styles to follow the standard fonts suggested by the HI alpha generations teachers. The sizes of the fonts indicated by the teachers of the HI alpha generations are title: 100 pt; instruction: 25 pt; button: 50 pt. Figure 1 shows the sample of the fonts.

Design Principle 6: The sizes and quality of the graphics, videos, and animations provided in the MSL mobile applications should be maximized and consistent.

The videos used in this prototype were linked with the school's YouTube channel, where the users can directly view the videos through this prototype. This software's limitation is that it does not allow the videos to be uploaded directly into this proto-type. Thus, the quality of the videos will depend on the users' Internet connections. Then, the images used for the exercises were captured directly from the model's hands and uploaded directly into this prototype. Thus, the quality of the photos could be maximized. Next, the cartoon characters were designed in the Canva online software and downloaded in a JPEG format. Hence, the images uploaded in this prototype were maximized in quality except for the videos, as it depends on the user's state.

Fig. 3 Pop-out dialogue

3.5 Usability Heuristics NMDG 5: Error Prevention

Design Principle 7: Timely feedback should be provided in the MSL mobile applications through the pop-out dialogue to catch the attention of the HI alpha generations.

To avoid the HI alpha generations from making massive mistakes, timely feedback was provided in this prototype through the pop-out dialogue. This pop-out dialogue was explicitly designed to attract the users' attention and notify their exercises' results. The pop-out dialogue with green color will notify them that they have answered the questions correctly and can choose whether they want to continue answering the following questions or go back to the previous ones. At the same time, the pop-out dialogue with red color was designed to notify that they had wrongly answered the questions. Minimal options were given to prevent them from making massive mistakes. Figure 3 shows the sample of the screen.

3.6 Usability Heuristics NMDG 6: Recognition Rather Than Recall

Design Principle 8: The button in the MSL mobile application should be clickable and consistent according to the functionalities.

The HI alpha generations depend entirely on visuals rather than plain texts. Hence, the teachers of the HI alpha generations suggested designing the button with the symbols. All buttons were designed with the arrow symbol in this prototype as shown in Fig. 3 to ensure that the users can recognize these buttons' functions at first glance.

Design Principle 9: The characters used in designing the MSL mobile applications should be recognizable and suitable to the contexts of the HI alpha generations.

In designing the interfaces of an application for the HI learners, particularly the alpha generations, their preferences should be considered to attract their interests. According to the teachers of the HI alpha generations, they are easily attracted and excited when they recognize something they used to see. In this prototype, the video learning was designed by providing the characters they used to see, such as the animals and spaceships, as shown in Fig. 2.

3.7   Usability Heuristics NMDG 7: Flexibility and Efficiency of Use

Design Principle 10: Ensure flexibility of navigation for the HI alpha generations in exploring the MSL mobile application.

In this prototype, the users were given flexibility in navigation as they could skip the questions they did not want to answer and miss watching the video learning. These two functions were provided to increase the satisfaction of the HI alpha generations. They need flexibility since they are easily affected by the emotions of something they were forced to do, as mentioned by their teachers. Figure 4 shows the sample of the screen.

Fig. 4 Efficiency and flexibility

3.8   Usability Heuristics NMDG 8: Aesthetics and Minimalist Design

Design Principle 11: Keep the interface design of the MSL mobile application simple and less crowded to minimize the cognitive ability among the HI alpha generations.

To minimize the cognitive load for the HI alpha generations, the academicians have suggested only having 5–7 elements per screen. This is to evoke their cognitive abilities in memorizing the elements provided on one screen.

Design Principle 12: The interface's color for the MSL mobile application should be minimal but attractive with a good color combination.

The earth tone color was chosen to design the applications suggested by the HI alpha generation teachers to provide a minimal touch and less crowded interfaces. Still, it could attract the attention of the HI alpha generations with the cartoon characters attached.

3.9   Usability Heuristics NMDG 9: Help Users Recognize and Recover from Errors

Design Principle 13: The MSL mobile application should provide the redo and undo button for the alpha generations to recover from their mistakes.

This prototype provides the redo and undo buttons associated with the pop-out dialogue for the HI alpha generations to recover from the mistakes. For example, they might mistakenly choose to exit the application, but they can recover from the errors by choosing the buttons provided. Figure 5 shows the sample of the screen.

3.10   Usability Heuristics NMDG 10: Help and Documentations

Design Principle 14: The manuals can be an option to be provided for the HI alpha generations to assist them in using the MSL mobile applications.

As suggested by the experts in the expert review activities, the manuals can be provided as an option for the HI alpha generations to refer to for help. The teachers of the HI alpha generations suggested providing the manuals through the video tutorial. This is because they quickly understand the instructions through the video rather than giving the manuals through the texts. Figure 6 shows the sample of the screen.

Fig. 5 Recover from errors

Fig. 6 Video tutorial for
manuals

3.11 Usability Heuristics NMDG 11: Selection Driven Commands

Design Principle 15: The click-and-play functions should be maximized in designing
the MSL mobile applications for HI alpha generations.

All the exercises provided click-and-play functions as recommended by the academicians and content experts to avoid massive user errors. They led to frustrations while experiencing the applications.

3.12 Usability Heuristics NMDG 12: Visual Representation

Design Principle 16: Use more multimedia elements than text and static images in designing the MSL mobile applications since the HI alpha generations rely more on graphics and visuals than plain texts.

The HI alpha generations are attracted more to multimedia elements or visuals such as graphics, video, and animation.

To conclude, the proposed design principles were successfully validated using the prototyping approach with the development of the MSL prototype. Hence, it was proven that the proposed design principles were reliable in designing and developing the MSL mobile application for the HI alpha generations. However, it is suggested to be validated through user experience testing with actual users to ensure the design principles and the prototype can meet their needs and satisfaction.

4 Conclusion

In a nutshell, the MSL mobile application prototype was designed to improve the mobile applications currently available and help HI learners learn sign language interactively. Additionally, a new set of design principles for the MSL mobile applications was put forth based on the latest NMDG to enhance the dependability and usability of the MSL mobile applications. Furthermore, it is essential to validate the proposed design principles to ensure they are trustworthy and may serve as a guide for future researchers in creating the MSL mobile applications for HI learners, especially the alpha generations. The success of designing and developing the MSL mobile application prototype demonstrates the success of the prototyping method in validating the proposed design principles for the MSL mobile applications for the HI alpha generations based on the NMDG.

Acknowledgements This research was supported by the Ministry of Higher Education (MoHE) of Malaysia through the Fundamental Research Grant Scheme for Research Acculturation of Early Career Researchers (RACER/1/2019/SS01/UUM//6), Universiti Utara Malaysia (UUM) and Sekolah Pendidikan Khas Sungai Petani, Kedah for cooperation.

References

1. A. Bussé, H. Hoeve, K. Nasserinejad et al., Prevalence of permanent neonatal hearing impairment: systematic review and Bayesian meta-analysis. Int. J. Audiol. **59**, 475–485 (2020). https://doi.org/10.1080/14992027.2020.1716087
2. B. Baglama, M. Haksiz, H. Uzunboylu, Technologies used in education of hearing impaired individuals. Int. J. Emerg. Technol. Learn. **13**, 53 (2018). https://doi.org/10.3991/ijet.v13i09.8303
3. K. Graydon, C. Waterworth, H. Miller, H. Gunasekera, Global burden of hearing impairment and ear disease. J. Laryngol. Otol. **133**, 18–25 (2018). https://doi.org/10.1017/s0022215118001275
4. B. Bala, L. Song, Android app for improvising sign language communication in English and Hausa. Int. J. Adv. Sci. **06**, 15–24 (2020). https://doi.org/10.31695/ijasre.2020.33687
5. N. Palfreyman, Sign language varieties of Indonesia. Sign Lang. Linguist. **20**, 135–145 (2017). https://doi.org/10.1075/sll.20.1.06pal
6. T. Siong, N. Nasir, F. Salleh, A mobile learning application for Malaysian sign language education. J. Phys. Conf. Ser. **1860**, 012004 (2021). https://doi.org/10.1088/1742-6596/1860/1/012004
7. N. Aziz, A. Omar, A. Mutalib et al., Uncovering the needs for a hybridized interaction design model for sign language learning through experts' feedback. J. Phys. Conf. Ser. **1529**, 042102 (2020). https://doi.org/10.1088/1742-6596/1529/4/042102
8. F. Jeddi, E. Nabovati, R. Bigham, R. Farrahi, Usability evaluation of a comprehensive national health information system: a heuristic evaluation. IMU **19**, 100332 (2020). https://doi.org/10.1016/j.imu.2020.100332
9. F. Paz, F.A. Paz, J. Antonio Pow-Sang, C. Collazos, A formal protocol to conduct usability heuristic evaluations in the context of the software development process. Int. J. Eng. **7**, 10 (2018). https://doi.org/10.14419/ijet.v7i2.28.12874
10. B. Kumar, M. Goundar, Usability heuristics for mobile learning applications. Educ. Inf. Technol. **24**, 1819–1833 (2019). https://doi.org/10.1007/s10639-019-09860-z
11. E. Melo, C. Primo, W. Romero et al., Construction and validation of a mobile application for development of nursing history and diagnosis. Rev. Bras. Enferm. (2020). https://doi.org/10.1590/0034-7167-2019-0674
12. S. Nouri, P. Avila-Garcia, A. Cemballi et al., Assessing mobile phone digital literacy and engagement in user-centered design in a diverse, safety-net population: mixed methods study. JMIR mHealth uHealth **7**, e14250 (2019). https://doi.org/10.2196/14250

Exploring IoT Security in IoT Devices

Amirul Azri Azmi, Zarina Mohd Hussin, and Suraya Mohammad

Abstract The rise of the fourth industry revolution (IR4.0) brought the rise of Internet of Things (IoT) along with it. However, one of the main pillars of IR 4.0, cybersecurity, was almost ignored. This is because cybersecurity often comes with additional implications such as extra cost which is not covered by the product price. However, if an incident is likely to occur, the damage done would be much bigger than the cost itself. This paper aims to identify the vulnerabilities inside the IoT application while understanding the level of threat a security breach inside the IoT poses. Previous papers reviewed indicate that IoT devices lack privacy and a lightweight security protocol. In this paper, the lack of privacy is tested in a simulation setting using GNS3 for packet sniffing, while the lightweight security protocol was tested using the denial-of-service tool on an Arduino Uno and NodeMCU ESP8266 board. The results achieved showed that there is a way for security to be improved.

Keywords IoT security · Arduino · NodeMCU

1 Introduction

The Internet of Things (IoT) is an assembly of interconnected devices, humans, objects, and services that can share data and communicate information, to achieve a goal in different sectors and applications. When the term Internet of Things was first coined, the question would be what could be defined as "things". Until recently, groups of researchers tried to propose a definition for the Internet of Things. In 2009,

A. A. Azmi · Z. Mohd Hussin · S. Mohammad (✉)
Advance Telecommunication Technology, British Malaysian Institute, Universiti Kuala Lumpur, Kuala Lumpur, Malaysia
e-mail: surayamohamad@unikl.edu.my

A. A. Azmi
e-mail: amirul.azmi07@s.unikl.edu.my

Z. Mohd Hussin
e-mail: zarinahussin@unikl.edu.my

A. Ismail et al. (eds.), *Advances in Technology Transfer Through IoT and IT Solutions*,
SpringerBriefs in Applied Sciences and Technology,
https://doi.org/10.1007/978-3-031-25178-8_11

Sarma and Girão defined the "things" from physical objects to virtual objects which exist with identities in the Internet connectivity [1]. Haller et al. [2] proposed IoT as "A world where physical objects are seamlessly integrated into the information network, and where the physical objects can become active participants in business process".

In Malaysia, as of 2012, there are 146 potential patents from Malaysian inventors which is viable to be licensed to investors to strengthen their competitiveness in IoT applications and services. The growth of IoT in Malaysia is supported by MIMOS with its strategies. MIMOS has analyzed the Malaysian demographics in order to assess the strengths and weaknesses of incorporating IoT. Among the advantages of applying IoT in Malaysia are the availability of various incentives such as tax exemption, high phone and Internet user penetration rate, and also well-established network operators which are able to provide stable connections [3]. With the rise of the fourth industrial revolution or commonly referred to as IR4.0, there is a significant increase in the IoT ecosystem as most factories and small businesses heading to automation as part of their operation, following two of the nine pillars in IR4.0 which is IoT and autonomous systems [4].

The IoT can interconnect devices, people, data, and processes, allowing them to communicate with each other seamlessly [5]. Thus, IoT can help to improve different processes to be more quantifiable and measurable by collecting and processing a large amount of data [6]. The IoT can potentially enhance the quality of life in different areas including medical services, smart cities, agriculture, water management, and many more [7]. Although the benefits are many, the IoT ecosystem also comes with several disadvantages such as privacy and security concerns, and rural adoption and adaption fear [3].

Thus, this paper aims to identify the vulnerabilities inside the Internet of Things application and show the level of threat a security breach inside IoT poses. In this paper, the statement is tested using the denial-of-service (DoS) attack on an Arduino and NodeMCU IoT system.

2 Literature Review

Cybersecurity is one of the pillars of Industrial Revolution 4.0 alongside IoT and cloud computing. While having more devices is great for the system, it increases the chance of a defective node being in the network. This section discusses the aspect of IoT security research in three sections. The first will be the general architecture of the IoT, the second is the security challenges in the IoT, and the last one will be future research and goals for IoT security.

2.1 General Architecture of IoT

There are three proposed general architectures of the IoT being proposed in the literature. Muhammad et al. [8] and Matharu et al. [9] proposed the design as four layers. The first layer is perception, consisting of data sensors, radio-frequency identification, and barcodes to identify each unique object in the network. The second is the network layer which transmits the gathered information obtained from the perception layer. The third layer is middleware; it takes the information for processing and takes action accordingly. The last layer, the application layer, realizes the application of IoT based on the needs of the user. In contrast, Mahmoud et al. [10], Abomhara and Køien [11], and Zhao and Ge [12] used three layers of architecture to determine the security challenges for the IoT. The three layers were the perception layer, network layer, and application layer. The roles of each layer are as defined previously but exclude the middleware layer.

In 2018, Yusof et al. [13] proposed redefined four layers of IoT architecture. The first layer is the things layer, and the second layer is the communication layer which replaces both the perception layer and the network layer as proposed in [9]. The third layer is the application layer which provides users control over the system. The last layer is the data analytics layer in which they examine and process the data to decide the best course of action.

2.2 Security Challenges in IoT

Zhang et al. [14] listed seven challenges for IoT security. Firstly, object identification. A new naming service is desirable which is lower computation and communication overhead. The second challenge is authentication and authorization in the network. It is currently not feasible to issue a certificate to each object since the total number of objects is normally huge. The third challenge is the privacy of the system, the nature of IoT systems sharing information always to each node in the network, and a way to protect the data in transport is needed. Fourth, the unavailability of lightweight cryptosystems and security protocols. It is limited by the computational capability of the hardware itself. The fifth challenge is software vulnerability and backdoor analysis. The sixth challenge is malware in IoT applications. The limited power in hardware makes real-time virus scanning not feasible. The last challenge is security issues from out-of-date Android operating systems which is largely data leakage.

On the other hand, Hossain et al. [15] presented the challenges based on several aspects of IoT which were information security, access level security, functional security, end device security, communication security, and service security. Information security requires the integrity of the data sent to guarantee no adjustments had been done, anonymity to hide the source of data, a node cannot deny a message it previously sent, and freshness to ensure no old message is being replayed. Access level security involves authentication, authorization, and access control; due to the low memory and processing power of IoT hardware, the protocol to authenticate and authorize each device is hard to implement. Function security involves exception handling when one node is compromised such as a denial of service, fails, and runs out of energy. End device security needs insecurity due to device category and capability, software, firmware, and storage security. Communication security challenges involve security in network service security and cryptographic security. Lastly, the service security requires native service security to prevent a lack of secure account credentials, and partner cloud service security which comprises third-party apps handling the IoT introducing an unknown threat.

Besides, Mahmoud et al. [10] listed security challenges based on the previous architecture layers. The perception layer has signals between the sensor node which can be compromised by disturbing the waves. The low storage capacity, low power consumption, and computation capability make them susceptible to denial-of-service attacks, eavesdropping, passive monitoring, and man-in-the-middle attack in the network layer. In the application layer, the large overhead of the application that analyzes data can have a big impact on the availability of the service. Another challenge will be to design an application based on how different users will interact with them and the amount of data that will be revealed. The last challenge listed was to create an ability to control what data is disclosed and be aware of how the data is used, by who, and when. Abomhara and Køien [11] listed the vulnerabilities of IoT devices. The first one is the Dolev-Yao (DY) model. The model when in effect in the network may intercept all or any message ever transmitted between IoT devices and hubs. Besides, IoT devices are vulnerable to denial of service since naturally, they have low computational power. Next, the devices are greatly vulnerable since they are naturally left in a place to work on their own which makes them prone to physical attacks. Lastly, attacks on privacy such as eavesdropping and passive monitoring where the message is not encrypted, and can be easily read by the interceptor, and traffic analysis combined with eavesdropping makes it easier to identify important and valuable information.

In another paper, Roman et al. [16] presented the first security threat as protocol and network security. IoT hardware lacks the resources to implement security protocols that lead to a gap in end-to-end security. Data and privacy threats can be exploited by entities that profile and track users without their consent. Lastly, the absence of fault tolerance in IoT. It is impossible to perform software patching for billions of devices. The object has to be secured by default and should be able to defend itself against network failures and attacks. Zhao and Ge [12] referred to the IoT architecture layer for their security challenges. In the perception layer, due to the nature of sensors being left alone to work, the signals are exposed in the public place. In the network layer, the current network security is not compatible with the system. It is designed from the perspective of a person. With the huge number of devices, if it uses the existing method of authentication, the large amount of data will likely block the network creating a denial of service on its own. In the application layer, the data access permission needs to be controlled using identity authentication to prevent illegal user intervention. The application should be able to deal with mass-data, as a huge amount of data transmission leads to network congestion which may lead to network interruption and data loss.

On the other hand, Matharu et al. [9] listed the first challenge in IoT security, which is its robustness in connectivity. An unstable connection to the Internet poses a major challenge. Next is interoperability and standardization of IoT devices. Devices made by various vendors differ in technologies and services making them incompatible with the same protocol to be implemented. Besides, data confidentiality and encryption techniques will be required to ensure the secure transfer of transmitted data and guard against unauthorized interference. Lastly, big data security is needed to ensure only relevant data is being extracted from huge databases.

2.3 Future Research and Goals for IoT Security

In 2015, Hossain et al. [15] suggested future research on IoT security should involve trust management which means being able to distinguish a node outside the network as trustworthy, and governance which is the amount of actual security control on the network of things. On the other hand, Mahmoud et al. [10] suggested four security countermeasures for IoT: First, authentication measures scheme based on hashing and feature extraction to avoid any collision attacks; second, trust establishment by creating a concept of mutual trust for inter-system security in IoT; third, having a federated architecture where every device follows a standard, which makes it easier to create a security protocol for it; fourth, raising the security awareness of device owners such as not using default passwords to enable hackers to conduct attacks against the whole network from one compromised device.

Furthermore, Mahmoud et al. [10] also presented future directions for IoT security. In the future, IoT devices should have architecture standards to enable better integration in the IoT framework. Next, devices should have a management entity that can monitor the connection process of devices to prevent identity theft. An abstract session layer should be accommodated as an additional layer in the IoT architecture that specifically manages connections, protocols, and sessions between heterogeneous devices. Lastly, newer IoT devices should consider incorporating the 5G protocol into the security measures.

Abomhara and Køien [11], presented criteria for IoT security measures. It needs to be lightweight and symmetric to support resource-constrained devices. A cryptographic technique that enables protected data to be stored, processed, and shared without the info leaking to other parties. Security and privacy issues should be considered seriously because IoT not only transmits sensitive data such as personal and business, but the devices also have influence over the physical environment with their control ability. Finally, new frameworks should be developed to address global ID schemes, identity management, identity encoding and encryption, authentication as well as the creation of global directory lookup and discovery services for IoT applications with various identifier schemes.

For future researches, Roman et al. [16] suggested cryptography and protocols research. A lightweight protocol fits perfectly with a low-resource device like IoT hardware. Next is an identity and ownership protocol plus privacy protection for data transmitted from the IoT. Zhao and Ge [12] also suggested putting access control on RFID tags and using data encryption on RFID signals. An intrusion detection technology to monitor the behavior of network nodes, while finding suspicious behavior of nodes should be added. The network capability of IoT devices should also be increased to consider the availability of the huge amount of data. Lastly, the application in IoT should be able to protect private information which includes fingerprint technology, digital watermarking, and anonymous authentication.

3   Methodology

Among the papers reviewed, a common security challenge that was presented was the low processing capability of the device making it not possible to implement the current method of security to the device. Because of that, the focus of this research is to investigate more on the issues by implementing the denial-of-service attack to demonstrate the low processing power of the IoT.

A denial-of-service (DoS) attack is an attack meant to shut down a machine or network, making it inaccessible to its intended users. DoS attacks accomplish this by flooding the target with traffic or sending it information that triggers a crash. Thus, the methodology implemented in this part of the attack will be by flooding packets into the port of the IoT board. As per the previous theory suggested by Abomhara and Køien [11] and others [10, 14–16], the device is vulnerable to denial-of-service attacks due to its low computational power. Thus, the objective of this is to study the effects of denial of service on the IoT board. Specifically, the parameter that will be measured is the throughput or the amount of data flowing through the network. The board that will be used in this test is the Arduino with ethernet shield WIZ5100 as the network interface and NodeMCU ESP8266. A comparative survey of both Arduino and NodeMCU as IoT devices is provided in [17].

The board will be a TCP server that is used by the Iperf server to measure its throughput. Iperf [18] is a tool that is widely used for network performance measurement and tuning. It needs no installation; the folder is copied to a destination which is later accessed through the command prompt window in Windows. Another tool that will be used in this test is Low Orbit Ion Cannon (LOIC) [19]. It is a tool used for simulating the denial-of-service attack.

The experiment setups for the Arduino and NodeMCU are shown in Fig. 1a, b, respectively. The experiment requires a router, an IoT board, and a laptop. In the case of the Arduino setup, the Arduino is connected to the router using a Cat5e Ethernet cable and so is the laptop. The Arduino is powered by the laptop's 5 V USB port. Figure 1a shows the connection between the board and the router. As for NodeMCU, since it has a wireless interface card, therefore, the connection to the router is wireless as shown in Fig. 1b.

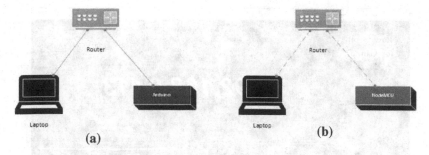

Fig. 1 Test setup for **a** Arduino board and **b** NodeMCU board

The steps for the performance test are as follow:

1. Set up the devices as in Fig. 1a for the Arduino performance test or as in Fig. 1b for the NodeMCU performance test.
2. Arduino IDE is launched from the laptop, and the board is selected with the appropriate port for the serial connection of the Arduino or NodeMCU.
3. The code for the TCP server for Iperf is uploaded to the board.
4. After the uploading finishes, run the serial monitor where it will show the network connection and the IP address for the Arduino or NodeMCU which will be typed later in Iperf.
5. Launch the command prompt by opening the Windows Start Menu and typing 'cmd'. Go to the iperf destination folder by using the change directory command: i.e.: cd c:/destination of iperf folder.
6. In the iperf subfolder, type in the command line "iperf -c 192.168.1.177 -t 100 -i 10 -w 300k" as shown in Fig. 2 where -t defines the total time of the test, we set as 100 s, -i refers to the interval of the data taken, we set as 10 s, and -w refers to the windows size of the packet transfer, we set as 300. The IP address is according to the Arduino or NodeMCU IP address in the serial monitor. Based on the command, the performance will be displayed on the command prompt in intervals of 10 s for a total of 100 s.
7. Start LOIC and in LOIC type in the IP address of the Arduino or NodeMCU and select the type of attack as TCP as shown in Fig. 3. LOIC denial-of-service application is now started and starts targeting the Arduino or NodeMCU to flood its network port.

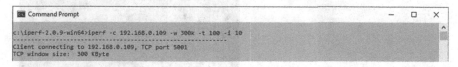

Fig. 2 Iperf command

Fig. 3 LOIC for Arduino and NodeMCU attack

4 Result and Discussion

Figure 4a shows the result from the Iperf test for the Arduino Uno showing a total of 9 Megabytes of data transferred with an average throughput of 746 Kilobits per second. This is the result before any denial-of-service attacks. Figure 4b shows the result from the Iperf test after packets start flooding the TCP port of the Arduino. The result showed data transferred equal to 9 Megabytes as before but with a slight decrease in throughput performance. The result for the NodeMCU test before packet flooding starts is shown in Fig. 5a. It shows the performance of NodeMCU with 5.12 Megabytes of total data transferred and 420 Kilobytes of data per second average. The result when the LOIC application starts flooding the NodeMCU TCP port is shown in Fig. 5. The attack caused the NodeMCU to crash and reset itself.

In this performance measurement test, two microprocessor boards were compared against each other to see whether each of them can withstand a denial-of-service (DoS) attack. The boards were the Arduino Uno with W5100 Ethernet shield as the network interface card, and another one is the ESP8266 NodeMCU with a built-in

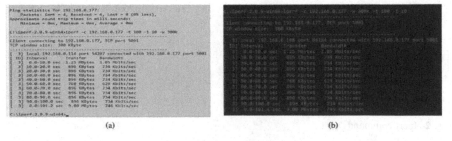

(a)   (b)

Fig. 4 Test result from Arduino Iperf test: (**a**) before the packet flooding and (**b**) after the packet flooding

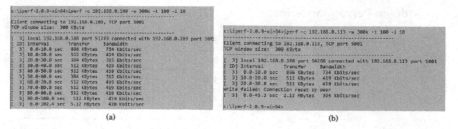

Fig. 5 Test result from NodeMCU Iperf test: (**a**) before the packet flooding and (**b**) after the packet flooding

wireless network interface card. The reasoning behind the chosen boards was that they were cheap and easily attainable. These points are exactly what hobbyists and beginners look for when trying to start a project. Also, the abundance of code and support available on the net makes it a very versatile board to be used for many situations.

The result showed that the Arduino withstands better against the packets flooding its IP address. On the other hand, the NodeMCU showed a significant performance decrease in its performance. The difference between the two boards in this test is the network interface card and the type of connection it uses. Both boards although running on different software can be accessed through the same programming code from the Arduino with a little library addition. The result shows that the Arduino was able to withstand better against the denial-of-service attack. The reason may be that it uses a wired connection to the router which provides a stable connection while the NodeMCU relies on the wireless connection which is affected by the signal strength. Although the test was done with the best signal strength available, the result was as shown above. The LOIC.exe crashed the NodeMCU and forces it to resets itself in the middle of the test. Furthermore, both tests were conducted 10 times, and the graph is displayed in Fig. 6a, b. In 2018, Bento et al. [20] did the same performance test on the NodeMCU using Linux as the denial-of-service application. The result of the test was the server that is on the NodeMCU that stops displaying its webpage. The result was similar to this test in terms of the denial-of-service crashing the NodeMCU making it reset itself.

5 Conclusion

In this paper, we demonstrate that vulnerabilities in the IoT do exist. The first vulnerability tested is the ease to intercept data from the signals of the device. A simulation was done to assess the level of threat that a sniffing attack using Wireshark can do to a device that is transmitting information through a network. Due to the lack of encryption methods inside the IoT device, this attack is very practical. It was later proven in the result that it can pose a major threat as it reveals the private IP address

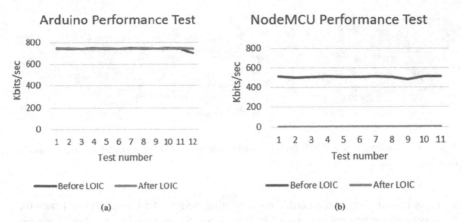

Fig. 6 **a** Arduino and **b** NodeMCU performance test comparison of before and after denial-of-service attack

of a device transmitting information which later can be used by the malicious person to specifically target a device such as an IP camera of the network. The second test done was abusing the low processing capability of the IoT devices. This is done by using a denial-of-service attack which is an attack that floods the device with packet requests thousands of times. The attack itself can disable the board and forces it to reset and reconnect to the network. A device reset can bring flaws to the data being recorded or in the worst case, the data is not synced to a cloud server or the data itself could be wiped from the smart device.

References

1. A. Sarma, J. Girão, Identities in the Future Internet of Things. Wirel. Pers. Commun. **49**, 353–363 (2009)
2. S. Haller, S. Karnouskos, C. Schroth, The Internet of Things in an enterprise context, in *Future Internet – FIS 2008. FIS 2008*, ed. by J. Domingue, D. Fensel, P. Traverso. Lecture Notes in Computer Science, vol. 5468 (2009), pp. 14–28
3. A.H. Abdul Halim, S.W. Yoong, M.S. Abdul Majed Shik, J. Hamzah, W.K. Goon, M.F. Amin, L. Sebastian, N. Jaafar, Z. Sayuti, N.Z. Musa, Z. Mohamad Nor, Y.H. Ng, National Internet of Things (IoT) strategic roadmap (2015), http://www.mimos.my/iot/National_IoT_Strategic_Roadmap_Book.pdf. Accessed 29 Aug 2022
4. C. Senn. The pillars of Industry 4.0 (2019), https://www.idashboards.com/blog/2019/07/31/the-pillars-of-industry-4-0/. Accessed 14 Jan 2021
5. N. Hossein Motlagh, M. Mohammadrezaei, J. Hunt et al., Internet of Things (IoT) and the energy sector. Energies **13**(2), 494 (2020)
6. F. Shrouf, J. Ordieres, G. Miragliotta, Smart factories in Industry 4.0: a review of the concept and of energy management approached in production based on the Internet of Things paradigm, in *IEEE International Conference on Industrial Engineering and Engineering Management* (2014), pp. 697–701
7. D. Bandyopadhyay, J. Sen, Internet of Things: applications and challenges in technology and standardization. Kluw. Commun. **58**(1), 49–69 (2011)

8. F. Muhammad, W. Anjum, K.S. Mazhar, A critical analysis on the security concerns of Internet of Things (IoT). Int. J. Comput. Appl. **111**(7), 1–6 (2015)
9. G.S. Matharu, P. Upadhyay, L. Chaudhary, The Internet of Things: challenges & security issues, in *IEEE International Conference on ICET* (2014), pp. 54–59
10. R. Mahmoud, T. Yousuf, F. Aloul et al., Internet of Things (IoT) security: current status, challenges and prospective measures, in *IEEE International Conference on ICITST* (2015), pp. 336–341
11. M. Abomhara, G.M. Køien, Security and privacy in the Internet of Things: current status and open issues, in *IEEE International Conference on PRISMSL* (2014), pp 1–8
12. K. Zhao, L. Ge, A survey on the Internet of Things security, in *Ninth IEEE International Conference on Computational Intelligence and Security* (2013), pp. 663–667
13. F.N. Yusof, M.I. Yahood, M.N.T. Salleh, Cybersecurity Malaysia Internet of Things (IoT) security framework, in e-Security, in *CyberSecurity Malaysia* (2019), pp. 78–80
14. Z.K. Zhang, C.Y. Cho, C.W. Wang et al., IoT security: ongoing challenges and research opportunities, in *7th IEEE International Conference on SERV* (2014), pp. 230–234
15. M.M. Hossain, M. Fotouhi, R. Hasan, Towards an analysis of security issues, challenges, and open problems in the Internet of Things, in *IEEE World Congress on Services* (2015), pp. 21–28
16. R. Roman, P. Najera, J. Lopez, Securing the Internet of Things. Computer **44**(9), 51–58 (2011)
17. A.C. Bento, IoT: NodeMCU 12e X Arduino Uno, results of an experimental and comparative survey. IJARCSMS **6**(1), 46–56 (2018)
18. iperf - the ultimate speed test tool for TCP, UDP and SCTP, https://iperf.fr/. Accessed 23 Mar 2023
19. Low orbit ion cannon (LOIC), https://www.imperva.com/learn/ddos/low-orbit-ion-cannon. Accessed 23 Mar 2021
20. A.C. Bento, E.M.L. da Silva, M. Galdino et al., An experiment with DDoS attack on NodeMCU12e devices for IoT with T50 Kali Linux. Int. J. Adv. Eng. Res. Sci. **6**(1), 18–24 (2019)

The Supply Chain Resilience of the Commercial Vehicle Business During the Implementation of the National Recovery Plan

Najwa Hannani Mohd Nasir, Nurul Arina Masrom, Najihah Roslizan,
Hairul Rizad Md Sapry, Jimisiah Jaafar, and Abd Rahman Ahmad

Abstract The COVID-19 pandemic has brought dramatic changes in many business sectors, particularly the commercial vehicle business, which before the COVID-19 pandemic has already experienced a slowdown due to uncertainty in global supply and demand. The introduction of the movement control order (PKP) and several other measures to curb the spread of the COVID-19 epidemic have exacerbated the efforts to recover from the current challenges. This study investigates the effect of COVID-19 disruption on supply chain resilience among commercial vehicle businesses in Malaysia. The results show that all variables significantly influence the supply chain resilience capability in the commercial business sector except for the risk management culture, which requires further validation. In the event of a disruption, the supply chain resilience capability of the organization is critical to absorb and adapt to the changes caused by the interruption and innovatively seek improvement in the operation to become stronger and prepared for any similar disruption in the future. The findings of this study provide valuable information to the business

N. H. Mohd Nasir · N. A. Masrom · N. Roslizan · H. R. Md Sapry (✉) · J. Jaafar
Industrial Logistics, Universiti Kuala Lumpur Malaysian Institute of Industrial Technology, 81750
Masai, Johor, Malaysia
e-mail: hairulrizad@unikl.edu.my

N. H. Mohd Nasir
e-mail: najwa.nasir03@s.unikl.edu.my

N. A. Masrom
e-mail: arina.masrom@s.unikl.edu.my

N. Roslizan
e-mail: najihah.roslizan@s.unikl.edu.my

J. Jaafar
e-mail: jimisiah@unikl.edu.my

A. R. Ahmad
Faculty of Information Technology and Management, Universiti Tun Hussein Onn, 86400 Parit
Raja, Johor, Malaysia
e-mail: arahman@uthm.edu.my

practitioner affected by the COVID-19 pandemic. It also enriches the knowledge of academicians for similar research in supply chain resilience field.

Keywords Supply chain resilience · National recovery plan (NRP) · Commercial vehicle business · COVID-19

1 Introduction

During the implementation of the movement control order (MCO) in Malaysia, most business sectors were closed to support the government's measures to curb the spreading of the COVID-19 virus into the community. It is also no exception for commercial vehicle business that continues to close even though the COVID-19 situation in Malaysia has improved, and many sectors are back to operating under a different phase of the Malaysian National Recovery Plan (NRP). The delay in resuming the business operation in this sector has affected the supply of new vehicles, parts, and components for other sectors, particularly logistics, which is one of the critical sectors that supported the government initiatives to accelerate the local and export economy requirements during the NRP.

Through the extensive literature review, four variables are identified and selected to develop a supply chain resilience model for this study. The variables are collaboration, agility, risk management culture, and re-engineering, which are used to test the direct impact on the supply chain resilience in the commercial vehicle business. The re-engineering variable is also tested as a mediator to understand its influence on other variables of the supply chain resilience capability. Seven hypotheses were developed to answer the impact of these variables on the supply chain resilience capability in the vehicle commercial.

2 Methodology

A total of 25 respondents who participated in this study came from various vehicle commercial business organizations. The respondents are selected among the managers that considered to be highly versed in the issues under investigation. The data obtained is deemed sufficient, according to [1], who suggested that research conducted on organizational representatives or top executives has a higher chance of obtaining a lower response rate. All data obtained are valid and used for further analysis using the partial least squares software (SmartPLS) 3.0, which is considered suitable for this study.

3 Results and Discussion

In this study, the outer model and the inner model assessments are used to examine the developed theoretical framework. The outer model analysis such as factor loadings, variance inflation factor (VIF), composite reliability (CR), and average variance extracted (AVE) are being investigated as shown in Tables 1 and 2. For the loading indicator, most of the constructs are more than 0.708, which meets the [2] suggestion, but only five sub-constructs are below 0.7. However, it is still acceptable for item reliability as they indicate that the construct explains more than 50% of the indicator's variance. For the variance inflation factor (VIF), all constructs achieved a moderately correlated VIF value in the 1–4 range, and multicollinearity is not a serious issue if the value of VIF is below 5 [2]. As for the composite reliability (CR), all constructs achieved a higher composite reliability value in the 0.70–0.90 range, which is satisfactory to good as suggested by [2]. For the Cronbach alpha, most of the constructs achieved values of more than 0.7, except for supply chain agility, supply chain design, and supply chain understanding at 0.617, 0.609, and 0.602, respectively. However, the low Cronbach's alpha value should not be interpreted as a less accurate measure of reliability, as items are not weighted based on construct indicator loadings as used by composite reliability indicators. In terms of average variance extracted (AVE), the values are more than 0.5, which is acceptable according to [2], except for supply chain design and supply chain understanding which get 0.481 and 0.460, respectively. The results show that all eight constructs (supply chain agility, supply chain collaboration, supply chain design, supply chain management culture, supply chain re-engineering, supply chain resilience, supply chain understanding, and supply chain supply base strategy) are valid measures that statistically are significant at $p < 0.05$. The last analysis in the evaluation of the external model is the discriminant validity of the construct according to the criteria set by [3], showing that there is no discriminant validity issue in this study.

The inner model analysis is used for examining the hypothesized relationships between the exogenous and the endogenous construct. The assessment quality of the model depends on its ability to predict the endogenous constructs, which in this study is between the developed constructs and readiness to adopt supply chain resilience. Meanwhile, the R^2 value shows that 72.9% of the variance in intention to adopt supply chain resilience is explained by the exogenous construct developed in this model. The value of R^2 indicated a moderate model according to [4].

Among the constructs (see Table 3), the model estimation showed there was no significant relationship between supply chain management culture and supply chain resilience (H4 was not supported). Other constructs supported the developed hypotheses (H1, H2, H3, H5, H6, and H7) which is ($t > 2$, $p < 0.05$).

The f^2 values of 0.020, 0.150, and 0.350 indicate that the predictor variable is a low, moderate, or strong effect in the structural model [5]. The result shows that the effect size for SC agility, SC collaboration, SC risk management culture, SC re-engineering, SC understanding, SC supply base strategy, and SC design on the SC

Table 1 Construct validity, dimensionality, reliability, and item (supply chain resilience)

Construct	Item	Loading	VIF	AVE	CR	CA	R2
Supply chain resilience	SCER3	0.827	1.572	0.664	0.878	0.815	0.729
	SCRE1	0.715	1.753				
	SCRE2	0.811	1.803				
	SCRE4	0.849	2.122				
Supply chain collaboration	SCMC1	0.856	1.830	0.628	0.835	0.707	
	SCMC2	0.764	1.701				
	SCMC3	0.754	1.191				
Supply chain agility	AGLT1	0.775	1.665	0.571	0.798	0.617	
	AGLT2	0.829	1.700				
	AGLT3	0.651	1.060				
Supply chain management culture	SCRMC1	0.907	1.769	0.675	0.861	0.780	
	SCRMC2	0.845	1.573				
	SCRMC3	0.699	1.569				

Table 2 Construct validity, dimensionality, reliability, and item (supply chain re-engineering)

Construct	Item	Loading	VIF	AVE	CR	CA	R2
Supply chain re-engineering	SRE1	0.736	1.299	0.651	0.848	0.731	0.385
	SRE2	0.813	1.557				
	SRE3	0.866	1.647				
Supply chain design	SCME2	0.636	3.942	0.481	0.733	0.609	
	SCME3	0.812	1.016				
	SCME4	0.617	3.911				
Supply chain understanding	SCME5	0.731	1.062	0.460	0.718	0.602	
	SCME6	0.661	1.215				
	SCME7	0.639	1.148				
Supply chain supply base strategy	SCME8	0.869	1.552	0.640	0.841	0.723	
	SCME9	0.818	1.700				
	SCME10	0.705	1.276				

resilience capability is 0.172, 0.054, 0.014, 0.235, 0.044, 0.078, and 0.237, respectively. Therefore, according to [6], the impact of independent variables (SC agility, SC collaboration, SC risk management culture, SC re-engineering, SC understanding, SC supply base strategy, and SC design) on SC resilience is a mix of moderate and weak effects on the value of R^2.

Table 3 Summary of hypothesis testing results

Hypothesis	Relationship	T-value	P values	Decision
H1	SCA ---> SCR	3.514	0.000	Supported
H2	SCC ---> SCR	2.053	0.041	Supported
H3	SCD ---> SCRE	4.548	0.000	Supported
H4	SCMC ---> SCR	1.062	0.289	Not supported
H5	SCRE ---> SCR	4.010	0.000	Supported
H6	SCU ---> SCRE	2.143	0.033	Supported
H7	SCSBS ---> SCRE	2.875	0.004	Supported

4 Conclusion

This study has tested three exogenous variables, and one mediate variable to investigate the resilience level of the vehicle commercial in Malaysia. The variables are supply chain collaboration, supply chain agility, supply chain management culture, and supply chain re-engineering that mediates supply chain design, supply chain supply-based strategy, and supply chain understanding. The results show that only risk management culture is not significant in influencing supply chain resilience, which differs from the previous findings that suggest the importance of developing the risk management culture in the organization. Two possibilities lead to the perception of the respondent in this study. Firstly, risk management is perceived as a tedious process involving data gathering for audit purposes. Not many organizations use risk assessment information to improve their business operation. As such, the respondents in this study may not be aware of the benefit of developing a risk management culture in the company. Secondly, the interaction within the supply chain network for vehicle commercials is leaner compared to the other sector that might have multiple complex tiers that are prone to disruption. It justifies the respondents' perception that there is no need to develop a risk management culture in the organization.

This study provides an avenue to understand the impact of disruption such as the COVID-19 pandemic on the supply chain resilience capability of vehicle commerce in Malaysia. By using the developed supply chain resilience model, the study managed to unveil several valuable pieces of information that are useful for the vehicle commercial business organization in developing the resilience capability in their organization [2].

Acknowledgements We would like to extend our sincerest gratitude to all the respondents who took part in this research.

References

1. Y. Baruch, B.C. Holtom, Survey response rate levels and trends in organizational research. Hum. Relat. **61**(8), 1139–1160 (2008)
2. J.F. Hair, G.T.M. Hult, C.M. Ringle et al., *A Primer on Partial Least Squares Structural Equation Modeling (PLS-SEM)*, 2nd edn. (Sage Publications Inc., Thousand Oaks, CA, 2017)
3. C. Fornell, D.F. Larcker, Evaluating structural equation models with unobservable variables and measurement error. J. Mark. Res. **18**(1), 39–50 (1981)
4. W.W. Chin, The partial least squares approach for structural equation modelling, in *Modern Methods for Business Research*, ed. by G.A. Marcoulides (Lawrence Erlbaum Associates Publishers, 1998)
5. J. Cohen, *Statistical Power Analysis for the Behavioral Sciences*, 2nd edn. (Lawrence Erlbaum Associates, Publishers, Hillsdale, NJ, 1988)
6. I.M. Ambe, The use of postponement decisions in determining supply chain strategies of light vehicle manufacturers in South Africa. J. Econ. Behav. Stud. **9**(3), 180–191 (2017)

Crosstalk Analysis for Different Types of PCB VIAS Models at High-Frequency Transmission

Siti Marwangi Mohamad Maharum, Nur Amirah Aseri, Rifhan Amira Mohd Razif, and Aimi Syamimi Ab Ghafar

Abstract Crosstalk among vias is one of the crucial signal integrity issues which would deteriorate signal quality in high-speed multilayer PCB at high-frequency operation. This research will compare crosstalk analysis in four-layered PCB among three different PCB vias models called Through-Hole Vias, Blind Vias and Buried Vias when they operate at 5–25 GHz. CST Studio Suite® is used to simulate the crosstalk issue in terms of its insertion loss. This paper will elaborate in detail on the interesting findings of the research.

Keywords Crosstalk · Printed circuit board (PCB) · Through-Hole Vias · Blind Vias · Buried Vias

1 Introduction

In high-speed propagation signals and the miniaturization of printed circuit board (PCB) size, the high performance of data and power transmission become an expectation to all designers, especially in multilayer PCB [1]. Vias which is a Latin word referring to paths or ways that had been introduced in PCB technology to allow multilayer signal connectivity from one layer to another layer such as the ground layers or

S. M. Mohamad Maharum (✉) · N. A. Aseri · R. A. Mohd Razif
Electronics Technology Section, Universiti Kuala Lumpur British Malaysian Institute, 53100 Gombak, Selangor, Malaysia
e-mail: sitimarwangi@unikl.edu.my

N. A. Aseri
e-mail: amirah.aseri@s.unikl.edu.my

R. A. Mohd Razif
e-mail: rifhan.razif04@s.unikl.edu.my

A. S. Ab Ghafar
Department of Electrical Engineering Technology, Faculty of Engineering Technology, Universiti Tun Hussein Onn Malaysia, 84600 Parit Raja, Johor, Malaysia
e-mail: aimi@uthm.edu.my

© The Author(s), under exclusive license to Springer Nature Switzerland AG 2023
A. Ismail et al. (eds.), *Advances in Technology Transfer Through IoT and IT Solutions*,
SpringerBriefs in Applied Sciences and Technology,
https://doi.org/10.1007/978-3-031-25178-8_13

power layers in PCB [2]. Technically, via which is abbreviated from vertical interconnect access, is implemented in PCB to ensure that there is a vertical electrical route between the multiple PCB layers [3]. However, there will be discontinuities on the PCB trace that will lead to signal integrity and electromagnetic interference issues, namely crosstalk. In PCB with the presence of vias, the crosstalk significantly occurs due to unwanted noise coupled from one via to adjacent vias, especially on a tightly pitched PCB [4, 8]. The crosstalk issue will become worse as the transmission frequency increases. This is because the signal resonates at certain frequencies based on the interconnect geometries and dielectric regions [5]. Besides that, design rules are very crucial to reduce the causes of crosstalk, i.e. the mutual capacitance and mutual inductance.

Hence, the crosstalk mitigation strategy becomes one of the major concerns among researchers and engineers in producing reliable PCB performance while taking into account the demands for high-speed data transfer, device miniaturization and high-density components usage. One of the methods to reduce crosstalk in the presence of vias is by using the circuit-concept approach for the transmission line section and impedance modeling for the via section. It was proposed in [6] in which the crosstalk had been significantly reduced through metal filled via holes within the bent transmission lines on a PCB. On the other hand, researchers in [7] succeeded to reduce the vias crosstalk without sacrificing the signal-to-ground ratio and maintaining the integrated circuit (IC) package ball map. Such achievements would not be simply achieved through straightforward methods such as the placement of orthogonal signal pairs, adding more ground vias nor increasing the gap between the vias. Meanwhile, far end crosstalk (FEXT) which is one of the crosstalk categories had been mitigated by cascading the via and pin area wiring structures [8].

However, in contrast to other published works, the focus of this paper is to conduct a crosstalk analysis in multilayer PCB when different PCB vias models are operated at high frequency. As listed in [9], there are eight types of vias models but only Through-Hole Vias, Blind Vias and Buried Vias will be covered in this paper.

2 Methodology

The methodology of this research is divided into two parts, namely (i) the design of three-dimension (3D) PCB vias models and (ii) the simulation setup of CST Studio Suite®.

2.1 Design of 3D PCB Vias Models

The scope of this paper is limited to four-layered PCB only. This is because four-layered PCB is the minimum multilayer PCB which can accommodate all three types of vias models, i.e. Through-Hole Vias, Blind Vias and Buried Vias on the same

PCB. Figure 1 provides an illustration aid for the above-mentioned vias models. Specifically, Through-Hole Vias is visibly seen on the PCB because the vias were drilled from the PCB's upper layer to the bottom layer. In this research scope, the upper layer and bottom layer will be referring to the first layer and fourth layer, respectively. Conversely, Blind Vias requires drilling depth accuracy because this vias model will provide signal interconnection from PCB's upper or bottom layer to any level of the internal PCB layer. Finally, Buried Vias could not be seen from any external PCB layer because the vias were drilled and electroplated between PCB's internal layers.

Figure 1 also acts as the design reference for all three PCB vias models because each PCB vias model will be then translated into its 3D simulation model in CST Studio Suite®. The length and width of each multi-layered PCB are set to 30 × 20 mm. Meanwhile, the cumulative thickness of all four-layered PCBs is set to 1.6 mm. The thickness measurement of every copper and dielectric for the designed four-layered PCB is shown in Fig. 2. These measurements were set based on the industrial recommendation although there are some measurements that might not be suitable for real implementation. Such limitation will be beyond the paper's scope because the aim of this investigation is to analyze the crosstalk behavior resulting from different PCB vias models.

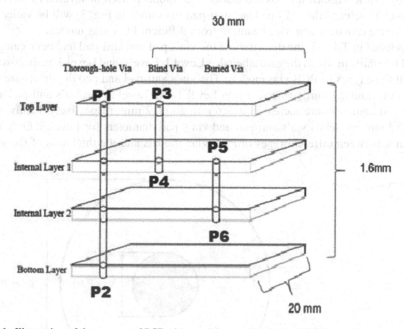

Fig. 1 Illustration of three types of PCB vias models on a four-layered PCB

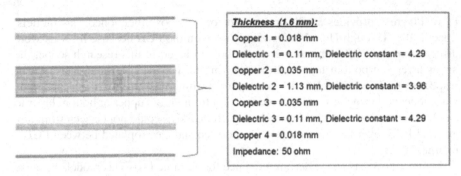

Thickness (1.6 mm):
Copper 1 = 0.018 mm
Dielectric 1 = 0.11 mm, Dielectric constant = 4.29
Copper 2 = 0.035 mm
Dielectric 2 = 1.13 mm, Dielectric constant = 3.96
Copper 3 = 0.035 mm
Dielectric 3 = 0.11 mm, Dielectric constant = 4.29
Copper 4 = 0.018 mm
Impedance: 50 ohm

Fig. 2 The thickness measurement of every copper and dielectric for the designed four-layered PCB

2.2 Simulation Setup of CST Studio Suite®

There are several specific product packages in CST Studio Suite®. This research uses CST Microwave Studio to conduct the simulation analysis on the targeted device under test (DUT). DUT for this research work is the 3D modeling which consists of Through-Hole Vias, Blind Vias and Buried Vias models on a four-layered PCB. Note that the diameter of the via's pad and anti-pad (as shown in Fig. 3) will be varied to analyze the crosstalk behavior resulting from different PCB vias models.

As listed in Table 1, the diameters of the via's pad and anti-pad had been categorized into three in which they are labeled as Level 1, Level 2 and Level 3, respectively, for all three types of PCB vias models. The via's anti-pad and pad diameters are set to 0.4 mm and 0.1 mm, respectively, in Level 1. For Level 2, the via's anti-pad and via's pad diameters are increased to 0.6 mm and 0.2 mm, respectively. Lastly, 0.8 and 0.3 mm are set as via's anti-pad and via's pad diameters for Level 3. Only the distance between different types of PCB vias models and the thickness of the vias

Fig. 3 Via's pad and via's anti-pad

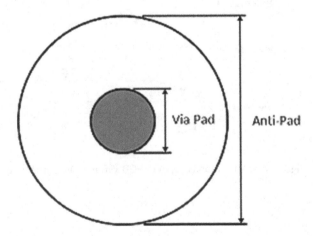

Table 1 Variation of via's anti-pad and via's pad measurements for Through-Hole Vias, Blind Vias and Buried Vias models

Types of vias	Through hole			Blind			Buried		
Level	1	2	3	1	2	3	1	2	3
Anti-pad diameter (mm)	0.4	0.6	0.8	0.4	0.6	0.8	0.4	0.6	0.8
Vias pad diameter (mm)	0.1	0.2	0.3	0.1	0.2	0.3	0.1	0.2	0.3
Thickness of vias (mm)	1.600			0.163			1.205		

for all, Level 1, Level 2 and Level 3, categories are set to constant values. In total, there are nine cases to be investigated. In Table 2, L1, L2 and L3 are denoted for Level 1, Level 2 and Level 3, respectively.

Next, all of these cases will be simulated in CST Microwave Studio using the time domain solver because it calculates the transmission of energy between various ports and/or open space of the investigated structure [10]. Besides that, the time domain could accommodate a wide frequency range of calculation through a single simulation [11]. Hence, this domain solver is a suitable method for this work since the frequency range of 5–25 GHz is the interest of the simulation investigation. The resulting S-parameter graphs will provide the return loss and insertion loss of the signal.

Figure 4 shows six defined ports denoted with roman number 1, 2, 3, 4, 5 and 6 on the targeted DUT. Based on the defined ports, each case (i.e. nine cases in total) will have six different graphs to demonstrate the insertion losses. However, among these six generated graphs per investigation case, there are only three important values of the S-parameter to be considered for further analysis. These parameters are expressed as follows:

Table 2 Nine measurement cases for DUT in simulation investigation

Factor	Anti-pad diameter			Via pad diameter		
Types of vias	Through hole	Blind	Buried	Through hole	Blind	Buried
Case 1	L1	L1	L1	L1	L1	L1
Case 2	L1	L2	L2	L1	L2	L2
Case 3	L1	L3	L3	L1	L3	L3
Case 4	L2	L1	L1	L2	L1	L1
Case 5	L2	L2	L2	L2	L2	L2
Case 6	L2	L3	L3	L2	L3	L3
Case 7	L3	L1	L1	L3	L1	L1
Case 8	L3	L2	L2	L3	L2	L2
Case 9	L3	L3	L3	L3	L3	L3

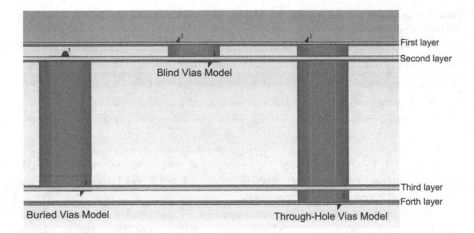

Fig. 4 Defined ports on the targeted DUT

$$S_{2,1} = \frac{\text{Transmitted signal power at Port2}}{\text{Incident signal power at Port1}} \tag{1}$$

$$S_{3,4} = \frac{\text{Transmitted signal power at Port3}}{\text{Incident signal power at Port4}} \tag{2}$$

$$S_{5,6} = \frac{\text{Transmitted signal power at Port5}}{\text{Incident signal power at Port6}} \tag{3}$$

3 Results and Discussion

In this research, the crosstalk behavior resulting from three different PCB vias models will be analyzed on a four-layered PCB from 5 to 25 GHz operation. To characterize such behavior, the via's pad and via's anti-pad measurements were varied based on the measurement values listed in Tables 1 and 2.

In total, 27 insertion loss graphs were generated from the CST Microwave Studio simulation from the 5–25 GHz range at Port 1, Port 3 and Port 5. In the 5–25 GHz range, the 15 GHz point was chosen for further comprehensive analysis of the generated crosstalk behavior. The obtained insertion loss values generated by these graphs were summarized in Tables 3, 4, and 5.

Based on Table 3, the highest insertion loss at 15 GHz for Through-Hole Vias model was during Case 7 with −4.585 dB, and the lowest insertion loss was produced during Case 2 with −3.497 dB. Meanwhile, in Table 4, the highest insertion loss at 15 GHz for the Blind Vias model was reported as −0.512 dB during Case 9, and the lowest insertion loss was −0.283 dB during Case 1. Referring to Table 5 on the other

Table 3 Insertion loss at 15 GHz for Through-Hole PCB Vias model

Different cases of study	$S_{2,1}$ parameter (dB)
Case 1	−3.499
Case 2	−3.497
Case 3	−3.521
Case 4	−4.049
Case 5	−4.028
Case 6	−4.025
Case 7	−4.585
Case 8	−4.552
Case 9	−4.519

Table 4 Insertion loss at 15 GHz for Blind PCB Vias model

Different cases of study	$S_{4,3}$ parameter
Case 1	−0.283
Case 2	−0.356
Case 3	−0.488
Case 4	−0.299
Case 5	−0.367
Case 6	−0.498
Case 7	−0.323
Case 8	−0.386
Case 9	−0.512

Table 5 Insertion loss at 15 GHz for Buried PCB Vias model

Different cases of study	$S_{6,5}$ parameter
Case 1	−2.331
Case 2	−3.293
Case 3	−5.172
Case 4	−2.316
Case 5	−3.258
Case 6	−5.151
Case 7	−2.317
Case 8	−3.247
Case 9	−5.415

hand, −5.415 dB during Case 9 was the highest insertion loss reported at 15 GHz for the Buried Vias model, while its lowest insertion loss was reported as −2.316 dB during Case 4.

From the above-reported values, Case 7 in the Through-Hole Vias model has the largest size of both via's pad and anti-pad, i.e. 0.3 mm and 0.8 mm, respectively. As compared to Case 2 in the Through-Hole Vias model, its insertion loss is the lowest because the sizes of both via's pad and anti-pad were set to the smallest size, i.e. 0.1 mm and 0.4 mm, respectively, in Case 2. The small sizes of the via's pad and anti-pad had created small parasitic capacitance and inductance values which are not severe as in Case 7. The same justification applies to crosstalk behavior in terms of insertion loss for Blind Vias and Buried Vias models. This proves that the via's pad and anti-pad diameters play a significant impact on the crosstalk. It can be said that these research findings answered the recommendation highlighted by the researchers in [8] stating that interconnect crosstalk can be minimized through fine adjustment on the via's anti-pad size.

Lastly, Fig. 5 demonstrates the comparison of insertion loss between nine simulated cases for $S_{2,1}$, $S_{3,4}$ and $S_{5,6}$ in Through-Hole, Blind and Buried PCB Vias models through a bar chart. Based on the overall insertion loss performance at 15 GHz, it is clear that the minimum insertion loss was generated by the Blind PCB Vias model. This is due to the shorter coupling length experienced by this vias model based on the simulated design shown in Fig. 4.

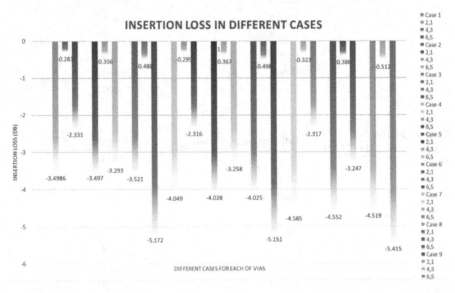

Fig. 5 Comparison of insertion loss between nine simulated cases for $S_{2,1}$, $S_{3,4}$ and $S_{5,6}$ in Through-Hole, Blind and Buried PCB Vias models

4 Conclusion

Vias are important in multilayer PCB layout as signals can be routed from one layer to another layer. In this research, crosstalk analysis in multilayer PCB when different PCB vias models called Through-Hole, Blind and Buried Vias models were evaluated at 5–25 GHz range on four-layered PCB. The crosstalk behavior was evaluated in terms of its insertion loss for nine cases for each PCB Vias model when the size of via's pad and via's anti-pad were varied to different measurement values. This research highlighted significant findings based on concerns raised in the literature and will be beneficial as a fundamental guideline for researchers and engineers who are working on the related PCB crosstalk issues.

Acknowledgements This research is funded by the Ministry of Higher Education (MoHE) Malaysia under the Fundamental Research Grant Scheme (FRGS) with reference code FRGS/1/2018/TK04/UNIKL/02/9. The authors would also like to thank the Universiti Kuala Lumpur British Malaysian Institute for the provision of research facilities.

References

1. L. Zhi, W. Qiang, S. Changsheng, Application of guard traces with vias in the RF PCB layout. EMC (2002). https://doi.org/10.1109/ELMAGC.2002.1177544
2. G. Dong, Y. Biao, D. Xidong, L. Yuan, Research on the influence of vias on signal transmission in multi-layer PCB. ICEMI (2017). https://doi.org/10.1109/ICEMI.2017.8265976
3. K. Hollaus, O. Biro, P. Caldera, G. Matzenauer, G. Paoli, G. Plieschnegger, Simulation of crosstalk on printed circuit boards by FDTD, FEM, and a circuit model. IEEE Trans. Magn. **44**(6), 1486–1489 (2008)
4. S. Wu, J. Fan, Investigation of crosstalk among vias. EMC (2009). https://doi.org/10.1109/ISEMC.2009.5284637
5. I.Y. Park, I. Ahmed, D. Brunker, P. Xie, J. Natarajan, Effects of various via patterns on resonance and crosstalk in high speed printed circuit boards. EMCSI (2020). https://doi.org/10.1109/EMCSI38923.2020.9191554
6. J.H. Kim, D.C. Park, A simple method of crosstalk reduction by metal filled via hole fence in bent transmission lines on PCBs. EMC Zur. (2006). https://doi.org/10.1109/EMCZUR.2006.214946
7. J. Wang, C. Xu, S. Zhong, S. Bai, J.J. Lee, D.H. Kim, Differential via designs for crosstalk reduction in high-speed PCBs. EMCSI (2020). https://doi.org/10.1109/EMCSI38923.2020.9191558
8. J. Tang, X. Yang, J.A. Hejase, M. Bohra, Y. Zhang, X. Duan, D. Kaller, W.D. Becker, D.M. Dreps, Far end crosstalk mitigation of differential high speed interconnects within printed circuit board via fields. EPEPS (2021). https://doi.org/10.1109/EPEPS51341.2021.9609151
9. PCBONLINE Team, 8 Types of PCB vias—an complete guide of PCB vias in 2021 (2021), https://www.pcbonline.com/blog/pcb-vias-and-their-types.html. Accessed 30 June 2022
10. C.W. Robertson, A.D.R. Phelps, C.G. Whyte, A.R. Young, K. Ronald, A.W. Cross, Simulations of waveguide components for use in a Ka-band gyro-traveling wave amplifier. ICIMW (2011). https://doi.org/10.1109/ICIMW.2010.5612605
11. K.K. Mistry, P.I. Lazaridis, Z.D. Zaharis et al., Time and frequency domain simulation, measurement and optimization of log-periodic antennas. Wirel. Pers. Commun. **107**, 771–783 (2019)

Printed in the United States
by Baker & Taylor Publisher Services

Printed in the United States
by Baker & Taylor Publisher Services